Acclaim for *Nature Noir* and Jordan Fisher Smith

"Daringly original and gorgeously nuanced . . . gives entrée into a strange, dark, and mesmerizing outdoor world that's absolutely unforgettable."
— *Boston Globe*

"Gracefully weaves scenes and stories with context, history, and reflection, in ways recalling the best of John McPhee." — *Los Angeles Times*

"Powerful with its intimate knowledge of place, *Nature Noir* achieves an even deeper mastery with its affection for the people and human histories of that place. Care and respect for a wild landscape attend to every page of this book." — **Rick Bass, author of** *The Lost Grizzlies*

"Taut drama . . . Smith's book follows the traditions of nature writers such as Emerson, Thoreau, John Muir, and Annie Dillard."
— *San Francisco Chronicle*

"Astonishing and fine." — *Cleveland Plain Dealer*

"A wonderful antidote to the treacly Ansel Adams image of our parks."
— *Wall Street Journal*

"Smith's is a refreshingly unsentimental kind of truth-telling."
— **Barry Lopez, author of** *Arctic Dreams*

"Smith tells engrossing stories [and] offers a fresh perspective on our threatened environment . . . *Nature Noir* reflects the spirit of an era as did *Desert Solitaire*." — *Charlotte Observer*

"Smith vigorously conveys the fascinating and often strange people and places he encountered." — *Tampa Tribune*

"By far the best book written by or about the modern park ranger I have read." — *Bloomsbury Review*

"Not only an electrifying tale of bringing the law to the wild west in the 1980s and '90s but also a graphic piece of writing from someone who has learned his craft from the royalty of American naturalists: writers like Gary Snyder, Aldo Leopold, and Edward Abbey." — *Buffalo News*

"It will cause readers to both thrill and shudder at the call of the wild."
— *Publishers Weekly*

"This book will tell you things you didn't know, and in a strong and original voice." — **Bill McKibben, author of** *The End of Nature*

"Eden had a snake, Yosemite Valley has a jail, but most nature writing is barricaded with omissions to make it just another gated community, one that Jordan Fisher Smith's powerful *Nature Noir* bursts open for readers. Thus in it he defends victims of domestic violence as much as violence against nature, which might not be separate things after all."
— **Rebecca Solnit, author of** *A Field Guide to Getting Lost*

NATURE NOIR

A Park Ranger's Patrol in the Sierra

JORDAN FISHER SMITH

A MARINER BOOK
HOUGHTON MIFFLIN COMPANY
BOSTON · NEW YORK

To Ray Stricker, Joe Burrascano, and Dan Abramson,
who rescued a rescuer.

First Mariner Books edition 2006

Copyright © 2005 by Jordan Fisher Smith

For information about permission to reproduce selections from
this book, write to Permissions, Houghton Mifflin Company,
215 Park Avenue South, New York, New York 10003.

Visit our Web site: www.houghtonmifflinbooks.com.

Library of Congress Cataloging-in-Publication Data
Fisher Smith, Jordan.
Nature noir : a park ranger's patrol in the Sierra / Jordan Fisher Smith.
p.   cm.
ISBN 0-618-22416-5
1. Fisher Smith, Jordan. 2. Park rangers—United States—Anecdotes.
3. United States. National Park Service—Officials and employees—
Anecdotes. 4. Natural history—Sierra Nevada (Calif. and Nev.)—
Anecdotes. I. Title.
SB481.6.F57A3 2005 363.28—dc22 2004059416

ISBN-13: 978-0-618-71195-6 (pbk.) ISBN-10: 0-618-71195-3 (pbk.)

Printed in the United States of America

Book design by Robert Overholtzer

QUM 10 9 8 7 6 5 4 3 2 1

This is a work of nonfiction based on the experiences of the author. However, some names,
places, physical descriptions, and other particulars have been changed. For that reason, readers
are cautioned that some details in the text may not correspond to real people, places, or events.

Chapter 7 of *Nature Noir* was previously published in a different form in the Autumn 1997 issue
of *Orion: People and Nature,* and in the anthology *Shadow Cat: Encountering the American
Mountain,* Susan Ewing and Elizabeth Grossman, eds., Sasquatch Books, Seattle, 1999. A passage
from the Prologue was previously published in a different form in the April 1996 issue of the
Wild Duck Review. Lyrics from "Don't Worry, Be Happy," written and performed by Bobby
McFerrin, © ProbNoblem Music, are used with the gracious permission of ProbNoblem Music.
Passages from *Lao Tzu: Tao Te Ching* by Ursula K. Le Guin, © 1997 reprinted by arrangement
with Shambhala Publications, Inc. Boston, www.shambhala.com.

CONTENTS

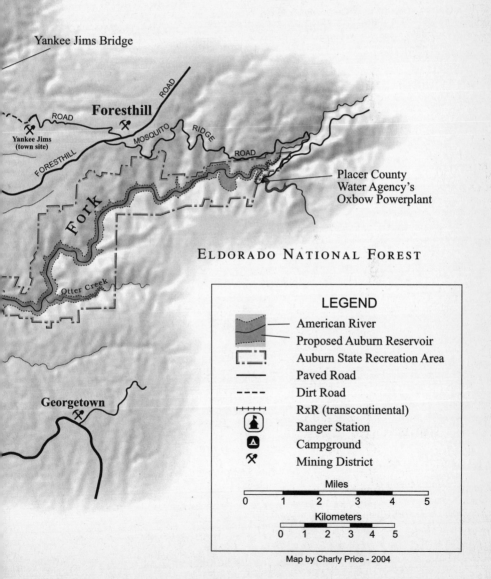

Auburn State Recreation Area
(Land owned by U.S. Bureau of Reclamation)

Iowa Hill

ROAD

Creek

Indian

TAHOE NATIONAL FOREST

Yankee Jims Bridge

ROAD

Foresthill

ROAD

Yankee Jims
(town site)

MOSQUITO

RIDGE

FORESTHILL

ROAD

Fork

Placer County
Water Agency's
Oxbow Powerplant

ELDORADO NATIONAL FOREST

Otter Creek

Georgetown

LEGEND

- American River
- Proposed Auburn Reservoir
- Auburn State Recreation Area
- Paved Road
- - - - Dirt Road
- RxR (transcontinental)
- Ranger Station
- Campground
- Mining District

Miles

| 0 | 1 | 2 | 3 | 4 | 5 |

Kilometers

| 0 | 1 | 2 | 3 | 4 | 5 |

Map by Charly Price - 2004

PROLOGUE

At the age of twenty-two, I decided to become a park ranger. I pursued that life with a freshness and single-mindedness I can scarcely bring to anything now. At twenty-eight, I could move around the mountains in summer or winter through any kind of weather. I could climb rock and ice, pack loads around the backcountry with mules, fix trails, build with logs, read maps and aerial photographs, and find my position in any kind of country. I was impervious to the sight of blood. I could splint broken bones and I could locate a large vein in an arm or leg by feel and get an intravenous needle into it to save a life. You could drop me into a small fire with another ranger and I could build line around it. You could drop me into a big fire with a crew and I could stay alive and sleep when the nights got long in the warm ashes in the heart of it. I could shoot a pistol and hit the target every time. I could go out on skis in the winter and live in snow caves. And I was, I thought, an accomplished lover of the land.

In my first summer with the Forest Service, I lived in an old fish hatchery near the southwest corner of Yellowstone National Park. From there I hiked up the Warm River in the orange light of evening through clouds of mosquitoes in the willow thickets around the beaver dams, startling sandhill cranes, which would burst from the brush with strange cries and circle over me in fear for their hidden nests. Later I lived in a tent in Granite Basin and then in Alaska Basin in the Grand Tetons, and I kept my pots and little stove in a hollow tree when I was away from camp. I lived at the end of the road in an alpine valley in Sequoia National Park, and in the autumn I returned to my cabin on horseback at night in the sleet with my feet so cold I couldn't feel them, to stand in front of the stone fireplace and drink hot tea. In the winters I would go back to a job on the coast of Northern California, where I went to sleep to the sound of the waves on the beach and the foghorns and bells on the buoys beyond the harbor. Still later I spent a summer living in a tiny cabin in western Alaska, a long airplane flight from the nearest road.

After all of that, I came to spend the greater part of my career watching over 48 miles of river and 42,000 acres of low-elevation California canyons that had suffered various forms of abuse for over a century and a half and had been condemned for a couple of decades to be inundated by a huge federal dam. And there probably wasn't a day when I didn't wonder how I came to choose this hopeless place on which to lavish my attention.

The path that led to the American River began like this: As a twenty-one-year-old student, I'd been spending every spare moment I could in the mountains and I'd resolved to find a way to make a living outdoors, as either a mountain guide or a ranger. As is often the case when you're young, in the end the choice between the two hinged upon happenstance. That September I sent out résumés for junior guide jobs — one I remember involved packing loads of supplies for the senior guides and their clients up the glaciers of Mount Rainier — but the season was over and no one was hiring, so late that month a friend and I decided to go climbing in Yosemite.

Leaving our car at Tioga Pass, we began hiking toward an ice climb in the cirque of Mount Dana, laden with rope, ice axes, and clanking equipment. As we made our way higher the forests grew sparse. Between patches of meadow and bare rock, the few remaining pines had been shaped by a hundred or two hard winters, the exposed grain of old lightning and avalanche wounds on their trunks weathered amber in the sun, their undulant canopies following the curves of the glacier-polished granite outcrops against which they'd huddled for protection from the wind. With a variety of tiny and efficient cook stoves available to backpackers and climbers, the Park Service prohibited campfires in these high forests so no one would use these works of art as fuel. But we came upon a campsite, and in it two men were chopping the heart out of one of these trees for an already overfed campfire. Greeting them with as much friendliness as we could, my friend and I tried to explain how long that tree had taken to grow. The two men sneered, told us to mind our own business, and went back to what they were doing. And as I watched them finish vandalizing that beautiful tree, I remember I wanted, more than anything, a ranger uniform and a citation book.

Less than three years later I had both, and I thought I knew what I was doing. I was a summer wilderness ranger for the Forest Service in what is now the Jedediah Smith Wilderness in Idaho and Wyoming. Unarmed and entirely untrained for police work, one day I happened to walk into a campsite full of people who didn't think much of my uniform or the government I represented. When the yelling and shoving were over I was physically intact, but the idea that people could always be dissuaded from breaking the law by a lecture or a small piece of pink paper with no immediate consequences was slowly dying in me. The following autumn, one of the rangers I worked with — a young woman who must have weighed all of 110 pounds — was threatened with an ax by drunken hunters in the Palisades Range. Still, I hadn't gotten into rangering to be a policeman, and it was another two and a half years before I summoned the gumption to enroll in my first law enforcement academy. When I got out, I took a job patrolling the high country of Sequoia and Kings Canyon National Parks. I felt strange and self-conscious at first wearing the gun and handcuffs, but I told myself they were mainly symbolic and I had no intention of using them on anyone.

One summer evening a couple of months later, an unarmed ranger on foot patrol in the mountains near where I was stationed found two men who had made an illegal campfire in the sparse whitebark pine forests at tree line, just as I had in Yosemite six years earlier. They'd been drinking, and when he confronted them one of them knocked him down, put both hands around his neck, and began to strangle him. Somehow the young ranger rolled out from under his assailant and called dispatch on his portable radio as he ran down a dark trail away from them. The dispatcher called me at my trailer in Silver City, and after talking with the ranger on the radio I sent someone to pick him up at the trailhead and telephoned one of my superiors at headquarters. I was the only law enforcement ranger in that part of the mountains that night, and when the district ranger heard what had happened, he arranged for a helicopter and another armed ranger to be sent up in the morning. The helicopter landed in the sagebrush on the floor of Mineral King Valley at first light, I boarded, and we flew east into the mountains. The rising sun was just gilding the highest peaks when we spotted the tent and the smoking remains of the campfire from the air. There was no sign of life; apparently our suspects were still sleeping. Our pilot set down on a patch of meadow several hundred yards downhill from the camp, and Ranger Roger Pattee and I crept up toward it through the rocks. When we got to the tent, we each chambered a round in our shotguns — which makes a loud, very serious-sounding metallic noise — and ordered the sleepy men out of the tent in their underwear. I was so green that once they were on the ground in front of us, I couldn't remember how to put on the handcuffs. But I managed. The victim subsequently identified his assailant, and within an hour and a half the two men were facing a federal magistrate, who was as unhappy about being dragged out of bed on a Sunday morning as they were.

In another three years I was working in the Guadalupe Dunes

along the central coast of California, and by then I had arrested all kinds of people. I had learned a couple of things about human nature that wouldn't startle you much if you took a moment to think about them: When regular people leave the city limits, their behavior doesn't change much, and habitual criminals are seldom rehabilitated by pretty scenery. Still, I believed there was one big distinction between me and your run-of-the-mill cop. I wasn't just slowing the inevitable decline of western civilization by arresting the guilty and carting off the wounded. I had been given a sacred charge: America's crown jewels, those special places legislatures had agreed were too good to ruin.

Driving through the rain and salt spray on those beaches in the winter of 1985, I felt terribly homesick for the mountains. But by that time jobs in the mountains had become difficult to get. With a new administration in Washington, the early 1980s saw much budget-cutting in parks. Park agencies were under pressure to fill what few openings they had with ethnic minorities and women, to correct their previous and very unfair practice of hiring exclusively white male rangers. That adjustment was long overdue, but its effect on my career wasn't good. And so it was in this climate of diminished expectations that I looked favorably on a position offered that winter by the California Department of Parks and Recreation: a patrol ranger job on the American River.

I had heard something about a large dam that for some time had threatened the American, but I wasn't planning on staying there anyway. It would only be a way station for me; I planned to transfer again within a year. So I applied for the job, my transfer was granted, and I reported for duty.

That a delegation of armed rangers was sent in to protect a piece of ground that could not be protected from the very government that employed them was an accident of history I will explain later. What I hadn't expected was how beautiful those canyons were. Or how I would come to both love and despise them, even fear them. There I witnessed some of the strangest incidents I've ever been involved in. What I will now relate of these events is entirely true, and in most cases taken from rangers' actual reports. Only the names and personal details of some of the rangers and other characters have been changed. In the case of the rangers who appear in these stories — and there were others who came and went over the years and whose names and stories could not be accommodated in a book of this length — that is appropriate. In their own way each of them did a good job under the toughest conditions imaginable, yet they never wanted to be known as heroes, and now they deserve their peace and quiet.

1 / A DAY IN THE PARK

IT WAS MIDSUMMER, a couple of years into my time in the foothills. A white haze filled the canyon of the North Fork of the American River, flattening its depth and dimensions. The heat was somnolent, the still air scented with the volatile spice of the brush fields.

The sides of the canyon were almost too steep to walk on and covered in thick stands of live oak, bay laurel, buckeye, and brush, the average color of which was gray-green. A sparse overstory of ghostly foothill pines cast patches of partial shade, their scabby gray trunks standing out from the canyon walls at precarious angles. At the bottom of the canyon a ribbon of blue water made its way southwest, bending back and forth across a broad gray bed of cobbles and sand. Half a mile downstream it entered a small, narrow reservoir, which followed the bends in the canyon for a couple more miles to where it spilled uselessly over a small dam. In the other direction, white anvilheads of cloud climbed over the high country, thirty miles east. There was a faint rumble of thunder. A few turkey vultures floated in intersecting circles along the canyon rim, savoring the hot air for the inevitable attrition of heat, drought, and violence.

A narrow dirt road descended the eastern wall of the canyon, winding in and out of the tributary ridges and gullies. Down it, a green Jeep station wagon with a police car's bar of red and blue lights and a couple of whip antennas bounced toward the river. A plume of dust boiled up behind the vehicle, then spread and settled into the surrounding woods.

Alone at the wheel of the Jeep was my sometime partner, Dave Finch. Today, as on every other day, the road meted out its daily increment of punishment to the Jeep's suspension and motor mounts and to Finch's lower spine and kidneys. From his right the two-way radio emitted the usual scratchy chorus of rangers and dispatchers outlining the progress of the repetitively mundane and gripping dramas of a summer weekend afternoon all the way down the river into the Central Valley. And as Finch fell deeper into the remove of the canyon, the voices on the radio became unintelligible bursts of static.

Three hundred feet above the river, Finch crossed an invisible line bisecting the canyon wall at a perfect level. It had been two and a half decades since Congress approved a dam that would flood this canyon. For over a decade the partially built project had been at a standstill, stalemated by politics, budgets, and the complexities of the very rocks under its footings. Yet it had never officially been called off. So everything below that line — the gray pines and olive-drab oaks, the wooly sunflowers on the road's cut-bank, the river, and even the little reservoir — could be seen only as temporary.

Finch came to the bottom of the road, and there he stopped and sat looking at the river, tanned elbow out the open window of the idling Jeep. In front of him, on the cobble beach that sloped away to the lake and the river that flowed into it, dozens of cars were parked haphazardly — economy cars, older pickups, works in progress with bald tires, spots of primer, and temporary registration stickers in their windows. Beyond them a small crowd of people lined the water's edge. Children splashed and squealed. Rough, tanned men — sheet-metal workers, drywall installers, meat cutters, heavy-equipment operators, electrician's apprentices, carpenters, unemployed truck drivers, occupants of trailer houses in the hills and cheap apartments in little foothill towns — stood along the beach with beers in their hands. Women — grocery clerks, dental hygienists, auto-part store delivery drivers with little butterflies tattooed on their breasts and little roses on their thighs — stood waist-deep in the water in bright-colored bikini tops and cut-off jeans, or lay on towels smoothing suntan oil on their skin, or hovered over their children, puffing on cigarettes and gesturing animatedly to each other. Big dogs — pit bulls, rottweilers, and retrievers — barked at each other on the beach and made forays into the water for balls and sticks. The rhythmic bass of competing car stereos and the squeals of the children and the barking of the dogs echoed off the far canyon wall.

Somewhere in front of Finch an engine revved. He paid no attention to it. Then there was a clatter of cobbles and the dull clunk of rocks hitting a car's underbody, a flash of sunlight on steel and glass, and his eyes fixed on an older American sedan with dull paint rapidly gathering speed toward him through the parked cars, its rear end fishtailing and its tires smoking and hurling stones.

The car began to straighten out as it accelerated up the cobble strand. From one side a man appeared, running hard toward it. He was wild-haired, stripped to the waist, his face contorted and mouth open in a yell, the big veins and fibers of his neck standing out. He clutched an object in the crook of one arm like a football player running a ball. The other hand was raised in a fist, which he waved angrily as he ran at the car. As the car passed him, he lifted the ball and extended his throwing arm, firing the ball perfectly toward the car. It flew through the open passenger-side window and disappeared inside the speeding vehicle. Except it wasn't a football, Finch thought. Something larger. A beach ball, perhaps.

Then Finch thought, *No, not a beach ball.* It was something pink, with limbs that moved as it flew through the air. *It's a baby that man just threw at that car. For the love of God, it's a baby,* he thought. Meanwhile the man continued to yell and shake his fist, running after the car, which still sped toward Finch.

Finch was not the only one whose attention was drawn to the commotion. As the baby sailed through the air, a low moan of horror and disbelief rose from the crowd on the beach, changing into a chorus of angry yells. As soon as they saw what he had done, several of the men and women down by the water rose to their feet and, unified by their hatred, began running toward the baby-thrower. As if in slow motion, Finch was spinning the Jeep around to position himself for a traffic stop and reaching for the controls of the light bar and siren and the radio microphone from its clip on the dash. The man saw the mob coming at him. Finch unfastened his seat belt. The car, passing by Finch's Jeep, braked to a stop. The driver's door flew open and a woman got out, screaming and waving her arms at Finch. Then she was reaching back into the car.

"Hey!" shouted Finch.

On the beach the mob descended upon the baby-thrower. Their aggregate intention now obvious to him, the man turned and ran at top speed down the beach along the lakeshore, disappearing into a thicket of the willow and giant rush. The woman had pulled the baby from the car and was holding it up in front of her, then hugging it, then holding it up again, screaming, weeping, red-faced, screaming something at Finch, pointing in the direction of the man who had run away. Finch could not tell what she was saying. He had the radio mike now, and in his other hand he gripped the aluminum club he carried in his Jeep.

"Hey!" he yelled again at the woman without knowing exactly what he wanted her to do, except to stop screaming so he could talk on the radio. Thumbing the microphone, he said into it, "Northern, four six nine, I need code three backup at Upper Lake Clementine."

There was a short crackle of static from the speaker in the car and then nothing. Finch keyed the mike again. "Northern, four six nine, code three traffic."

"Four six nine, Northern," the dispatcher replied, sounding almost bored. "Please repeat your traffic."

"Northern, four six nine. I need additional units to a four-fifteen at Upper Lake Clementine."

"Additional units, Lake Clementine . . ."

Finch imagined his backup roaring to the wrong end of the lake. "Negative, Northern — *Upper Lake, Upper Lake!*"

There was another burst of static. Then silence.

Finch: "Did you copy, Northern?"

Deafening static, then the dispatcher: "Copy, Upper Clementine, units available, please respond."

"One seven nine," I said into my microphone, several miles away. "Code three from the Confluence."

If the world exists in a perpetual state of uncertainty, if things are half-assed and watered-down and most things fall into a gray area, when you respond to a call like that you are bathed for a few minutes in superhuman certainty. You put away whatever squabbles you and your partners have had, ready to wade into the fray, to sacrifice yourself for any one of them. You hit the lights and siren and drive better than you normally do, think sharper than you normally do. The people in other cars look at you as you pass them on a mountain road and at intersections the cars part for you like the Red Sea for Moses. It is an acceptable substitute for reality; it's fleeting, but it keeps you believing in what you do.

One after another, three other four-wheel-drive patrol trucks converged on the road and roared down it, arriving at the bottom with the brakes stinking and spongy under the pedal. I was a couple of minutes ahead of the others. I jerked my baton from where I kept it jammed between the seat back and the cushion and rolled out of the car, sliding the club into the ring on my gun belt as I strode through the crowd now milling around Finch — fifty, perhaps seventy-five people.

Finch, poker-faced and sturdy in his green jeans, khaki shirt, gun belt, and green baseball cap with a badge insignia above the bill, stood by the open door of his car with the woman. Her car was still parked where she had stopped it, the driver's door still open. Finch was asking her questions and taking notes on a clipboard. As I walked up to them he glanced at me and, without acknowledgment or greeting, began to speak with no trace of excitement other than the elevated volume of his voice and the pace of his delivery.

"This was a male-female fight. The guy — he's gone — ran downstream. They were arguing. She says when they decided to separate, one was going to leave with the car and the other would stay. Then there was more arguing about who got to take the car and who had to stay with the baby. She jumps in the car and tries to leave, and he runs after her. That's when I saw him throw the baby at her in the car. I thought it was a beach ball. Then I thought, *Shit — it's a baby.* Luckily it passed right in through the window."

"Where's the baby now?" I asked him.

He pointed to the shade of the willow trees next to us. "Those two women in the crowd offered to hold it."

I looked over at the trees. The baby was naked but for a paper diaper, face flushed, in the arms of a woman forty feet from where we were standing. She and another woman were making worried-looking ministrations over it.

"Is it okay? You wanna call an ambulance?" I asked.

"Seems okay. I've called Child Protective Services to pick the little guy up and have him checked at the hospital. Anyway — then the crowd turned on the man and I thought they were going to kill him. The guy saw what was about to happen to him and ran into the brush down there." Finch gestured toward the thickets by the beach. "That was a good twenty minutes ago, and I have no idea where he is by now."

As Finch finished his account, the other rangers arrived, rumbling along the road in clouds of dust. Finch went back to questioning the woman. I walked over to where the others were getting out of their rigs. They were surrounded by bystanders who wanted to tell them what had happened and demand that something be done about it. When I told them what Finch had told me, the rangers were only too happy to leave their petitioners and search for the missing suspect.

The way it worked with us, as soon as the adrenal part was over, someone would have to pay for all the fun. You paid by having to write the whole thing up, a process that could take an hour of note-taking in the field and several hours to a couple of days back at the ranger station. As a rule, the first ranger on the scene was the one who paid. You labored over your account of the incident, all the while knowing that the DA would flush most of the nefarious acts you described down the drain and deal the guy out on a felony specified as a misdemeanor. At sentencing, the judge would impose a suspended sentence because the jail was full, or maybe once he was out on bail the guy wouldn't bother to show up for his arraign-ment. A bench warrant would be issued and when he got picked up a year and a half later on that and the seven other warrants he'd ac-cumulated by then, expediency dictated that all his cases be bound up and sold at a discount, and your charges might not even make the cut. So in the end he'd do a little jail time on some unrelated beef and no one would ever know what a beautiful job you'd done on the investigation. Year after year you wrote up these stories, and they'd wind up archived in a pile of cardboard boxes in the ware-house, flattening and drying like pressed flowers under the weight of all the stories above them — the unknown stratigraphy of your career.

In this case it was Finch who got to cut paper. To assist him while he continued taking the woman's statement, I began circulating to talk to the witnesses. The sweat ran down my face and fell in big brown dusty drops from my nose, staining my notes. My ballpoint refused to write on the wet spots. Our radios crackled with inartic-ulate static from Folsom Lake. The bystanders began to drift away, back down to the cool water.

It went on like this for a while. The whole affair had the usual combination of gripping danger and utter senselessness. Then I heard Finch on the little speaker-mike from the radio on my gun belt, clipped to the epaulet of my shirt: "One seven nine . . . that's the guy — long hair, no shirt — coming toward us."

I looked at Finch. He was pointing to a lanky man with unkempt hair walking up the sandy track from the willow thickets. The remaining spectators around us began to yell: "That's him! Aren't you going to do anything? That's the guy who tried to kill her baby!" I took a few steps toward the man, placing myself between him and the angry bystanders. He wore only dirty athletic shoes and a pair of cut-off jeans. He looked dazed.

"Put your hands up," I commanded him, pulling my baton from its ring. I didn't brandish it. Instead, cocking my wrist, I aligned it along the back of my forearm, where it wasn't threatening but was instantly ready.

"Turn around. Interlace your fingers and put your hands behind your head. Spread your legs. Don't move." I stepped around behind the man and patted the pockets of his shorts for weapons. Then I handcuffed him and, leading him over to my Jeep, put him in the back seat behind the expanded-metal prisoner cage. His sweaty back made a muddy smear across the dusty vinyl of the seat back. He looked weary. He said nothing and avoided my eyes. I didn't question him. He wasn't going anywhere, and now that he was captive, he had to be read his rights. I was more eager to question the witnesses on the beach before they disappeared, so I left him in the Jeep with the air conditioner on full blast and we went back to work on our notes and interviews. A breeze off the river stirred the leaves on the willow trees and momentarily cooled me, blowing through the sweat-soaked shirt on my belly, below my bulletproof vest. This thing was pretty well over.

Twenty minutes later I had my part of the statement-taking done. I returned to my Jeep to drop off a page full of notes and get a drink of water. Glancing at the prisoner in the back, I saw him slumped over sideways. I took off my sunglasses and studied his face. It was blue, ashen blue, like a dead man's. He was absolutely motionless.

"Finch! Look at this!"

Finch walked over and peered at the man through the side window of the car.

"He's faking it," he said.

"The hell he is."

"He's faking it."

"I don't know how he could fake that color. Take a look."

I opened the door, leaned into the back seat, and put my hand on the man's clammy chest, feeling for movement. Holding my face cautiously close to his, I listened for breathing. "Nothing," I told Finch over my shoulder. "He's not breathing."

"Shit," said Finch.

I reached for the latch on the man's seat belt. Grabbing his feet, I dragged them up off the floor. He was dead weight. I pulled on his legs. Finch shoved in next to me and grabbed one foot. The man tumbled out of the car onto his back on the rocky beach.

Kneeling on the rocks next to the still body, I rolled the tips of the index and middle fingers of my right hand down from the prominence of the man's Adam's apple to the carotid artery, feeling for a pulse. Finch was on the radio calling for an ambulance.

"His heart is still beating," I told Finch.

I jumped up and ran around to the back of the Jeep, opened the tailgate, opened the equipment box inside, and jerked out the medic's pack and oxygen kit. I ran back around the Jeep, put them down, ripped open their cases, and cranked open the oxygen supply valve. The regulator made a reassuring hiss as the gauge spiked up. I pulled on a pair of surgical gloves. I reached into the medic's kit for an airway, sized it against the man's clammy jaw, discarded it for another, opened his mouth gingerly with a finger and thumb, and threaded the curved plastic tube over his blue tongue and down his throat. Finch was uncoiling the shiny green supply line for the demand valve and handed the valve to me. I hit the button once: It made a satisfying *shush*. I picked up a mask from the kit and press-fitted it onto the valve, pushed it over the man's mouth and nose, and began to breathe him. Dispatch called; our ambu-

lance was en route. I reached for the speaker-mike and acknowl-
edged their transmission.

For maybe half an hour, maybe forty minutes, I watched his
chest rise and fall in response to the oxygen I forced into it with the
button under my thumb. Periodically I'd stop to check for a pulse.
His heart was still beating. Weak, but beating. With Finch and
the other rangers to keep an eye on the crowd, my world got very
small and simple, just the *sshhush* of the demand valve, the still
body, and the rounded river rocks beneath it.

Around the body were cobbles of greenstone the color of jade,
and granite ones with sparkling salt-and-pepper crystals. There
were river-rounded schists, the alternating layers of black and white
minerals across their flanks like stripes on a zebra. There were charcoal-gray gabbros. There were tan quartzites in which more wear-
resistant veins of quartz stood out in bas-relief, branching like the
blue veins on the still man's pale arms. There were eggs of porphyry
the color of dried blood and orbs of milky quartz blasted by nineteenth-century gold miners from fossil riverbeds high on the canyon walls upstream, where they'd lain entombed for fifty million
years since those rivers had been buried by volcanic eruptions.
Back in the living world now, these stones were orphans, because
the mountains from which those ancient rivers had plucked them
had long ago been washed down to the sea. Each rock and its tex-
ture, each lungful of oxygen, each moment, and then each next moment — these are all life is made of when nothing else can be
counted on. And for this reason there is a strange peacefulness at
the center of catastrophe.

After a while, the man's face began to pink up. His limbs twitched.
The airway I'd put down his throat began to bob and click against
the interior of the clear plastic oxygen mask. He was coming to, and
as he did, his gag reflex was coming back. Quickly I lifted the mask
and pulled the airway out of him so it wouldn't cause him to vomit
and inhale his stomach contents, which could lead to pneumonia

that might kill him slowly later, if he didn't die before the ambulance got there. His eyes fluttered. He took a couple of ragged breaths, and then another. Then there was nothing. Then another breath. Then nothing.

He had stopped breathing again. Again I inserted the airway and began moving his air for him. It went on like this two, then three times.

One of the other rangers stood over us, watching. The mob had gathered in a circle around us. "Is he dead?" a woman asked. "I hope so," some guy answered.

There were needle tracks on his arms. When he had run away, he must have gone in the bushes and fixed himself up with a speed-ball — a heroin and methamphetamine cocktail.

"Where's the damn ambulance?" I asked Finch, watching the man's chest deflate for the umpteenth time and glancing at the declining pressure gauge on my O_2 tank. *It'll be harder to keep him alive if I run out of oxygen,* I thought. I heard Finch calling dispatch for a status on the ambulance.

Eventually the ambulance got there. The other rangers moved the crowd of bystanders out of the way. A man and a woman in dark blue jumpsuits took over my patient, placing him on a gurney while I continued breathing him. Then, when they were ready, I pulled my mask off him and they replaced it with theirs. We exchanged paperwork rapidly, and they loaded and went up the road, their amber and red lights blinking through the trail of dust behind them.

I stood with my hands on my aching lower back, arched backward to stretch. My knees were sore from the rocks, a thing I hadn't noticed until now. I looked over at Finch and grinned, shaking my head. "Faking it, huh?"

"Yeah, well . . ." He shrugged his shoulders, grinned. I shrugged, grinned back.

With the adrenaline wearing off came the weariness, the dry mouth, the hunger. I drank a quart of water from the Jeep. I picked up my medical kit, equipment, bits of gauze, and green rubber surgical gloves off the rocks, tried to dust off my green jeans, found a bandanna and wiped the muddy sweat from my face. In a few minutes I heard the ambulance hit the Foresthill Road, where its siren came on. The wail echoed off the canyon walls above us for a period of minutes, then grew fainter and trailed away down the Foresthill Divide.

Back at our ranger station fifty feet below the waterline of the Auburn Dam in the lower North Fork canyon, I let myself into the front room that had once been the kitchen of the firefighters' mess and now functioned as our combination locker room, lunchroom, and secondary office. I flicked on the switch by the door. The cool fluorescents blinked and buzzed to life. I slumped into one of the old oak chairs around the big table in the center of the room, kicked my feet up on the table, reached for the phone, and dialed the number for the ER at the little hospital in Auburn. The line rang and I flipped open my lunchbox, unwrapped a sandwich, and took a bite. A nurse answered the phone. I told her I wanted to check on a patient we had sent in and I gave her the man's name.

"I'll let you talk to the doctor about that. He's right here," she said. She put the phone on hold.

I took another bite of the sandwich, leafing distractedly through a stack of wanted-fugitive bulletins and be-on-the-lookouts on the table.

The doctor came on: a guy I knew. I told him I was calling to see if my man made it.

"Yeah, he's fine. It was an overdose. We're running bloods, but I'd say from the agitated behavior followed by the loss of interest in breathing it's probably some mixture of heroin and a stimulant like cocaine or crank. Anyway, from what the medics said, you guys did a great job —"

"Oh —"

"— and I got a little from them about what our guy had done before he coded, you know? So it looks like you've saved his miserable life. I guess that should make you happy."

I thanked him and hung up, took another bite of the sandwich, leaned back in the chair, and stared up at the pale yellow paint on the pine planks of the ceiling.

"There are no innocent victims in this place," Finch always said as we watched the same people appear in alternating roles over the years. One day your guy was a perpetrator; a week or a year later he was a victim. Five years and a couple more tattoos later, you arrested him again as a perpetrator. Eventually he might wind up dead, drowned in the river or killed in a car crash or shot by one of his peers, and you listed him in the blank on the report where it said "victim."

The exception was an innocent like the Beach Ball Baby, as Finch was to call him from time to time when we would recount the story over the gales of laughter that were always our substitute for ennui. Then again, by now that little boy must be well into high school, and if his life turned out as badly as it began, he may already have qualified for a juvenile offender record, an obituary, or both.

But I like to think not. I like to think he got lucky, got placed with foster parents who loved him and lavished the good things on him. Perhaps he'll be valedictorian of his senior class and grow up to be a teacher, social worker, political reformer — who knows, maybe even a ranger.

I am less sanguine about his father's prospects. By saving him I set him loose again upon the world and, God help me, perhaps upon that little boy, unless of course the courts did their job — and when could that ever be counted on? But you never know. Perhaps there was some purpose served by that man's survival, some good he would do later to redeem himself. By the time of the Beach Ball Baby I was beginning to tell myself things like that. In any case, a park ranger is a protector. You protect the land from the people, the people from the land, the people from each other, and the people from themselves. It's what you are trained to do without even thinking, a reflexive and unconditional act. If you're lucky, you get assigned to people who seem worth saving and land and waters whose situation is not hopeless. If not, you save them anyway. And maybe in time, saving them will make them worth it.

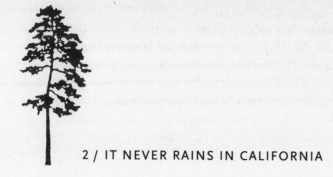

2 / IT NEVER RAINS IN CALIFORNIA

IN THE LONG SUMMERS, we were men of dust. We made our wages, our car and mortgage payments, our retirement plans, and our medical and dental benefits in the dust. We started out as young men and slowly went gray in the dust. In their late forties the older rangers began to need reading glasses to count the crumpled dollar bills they collected in our makeshift campgrounds, and the glasses would get dusty too, in their leatherette cases in the breast pockets of the rangers' uniforms.

On hot afternoons the dust mixed with the sweat on your face, and when you looked down to make a note on your clipboard, a drip from your brow streaked the inside of your sunglasses, blurring your vision. So you carried a couple of bandannas and wiped your sunglasses constantly, and soon the glasses were covered with tiny circular scratches that sparkled in the sunlight, for the dust was a competent abrasive.

The dust got into the thumb breaks of our holsters, and pretty soon our guns wouldn't come out. It found its way deep into the works of our pistols, where it combined with gun oil to form a brown deposit on the sear release levers, recoil spring guides, firing-pin springs, and decocking levers. We would fieldstrip and clean our weapons, but the dust went deeper than fieldstripping could reach.

Once a year the department's armorer would show up at our ranger station to inspect our guns. One by one, he'd disassemble them, and you could see him shaking his head and lifting his elbows as he applied himself to cleaning out the brown gunk. After a couple of days of this, still shaking his head, he would depart. And in the aftermath of his visits none of us ever received one of those congratulatory memos from headquarters for exemplary care of our equipment, as rangers did in those tidy little nature parks with neat little campgrounds on paved roads and no bullet holes in the signs.

At the district maintenance shops, mechanics looked up wearily on the approach of one of our squeaky Jeeps or Jimmies. The dust clogged air filters; it made a mockery of seals; it sullied lubricants and wore out bearings. It made our spotlights narcoleptic; alone after midnight on a vehicle stop, you'd twist the interior handle and flick on the switch and the light would remain asleep, face-down on the hood of your car.

The dust conspired with the rough roads to gain entry into the farthest reaches of our vehicles' interiors. After thirty or forty thousand miles, each new patrol wagon began its complaint of rattles and squeaks, and as its tailgate loosened and the dust worked its way past the gaskets, you'd see it rising behind you like smoke in your rear-view mirror and taste the grit on your teeth. Once inside, it made its way through the cracks around the doors of the plywood equipment cabinets behind the back seats and then crept through the zippers of our rescue packs, where it polluted sterile bandages, trauma shears, stethoscopes, oxygen masks, and whole sets of airways.

And so in my first summer in the Sierra Nevada foothills I learned to swathe my rescue gear in clear plastic trash can liners from the park maintenance shop. The wide, flat brim of my ranger hat served admirably to shield my face from the searing sun, but our cowboy-cut uniform jeans bound my sweaty legs so I could barely walk uphill. I learned to buy a kind of green jeans that looked official but were loose in the legs and seat, and to cut the label off the back pocket so my subterfuge was not detected. I began carrying at least two quarts of water and drinking it constantly. I learned to breathe through my nose and acquired an obscure yogic trick of drawing water into my nostrils from my cupped hands several times a day and forcibly expelling it to blow the mud out.

I learned that heat and dust made people testy, so I kept a tight leash on myself when things got tense. But there was little I could do to control their effect on others, and a good many of the men I met in the American River canyons carried guns (in case of rattlesnakes, they always told me), so I wore my bulletproof vest every day. The vest retained heat in the worst way, but I learned to discipline my mind so as not to be panicked by the claustrophobic discomfort of it. Whenever I had to talk to someone or take a report, I learned to walk to the shade of an overhanging tree and beckon to my reporting party to join me there. By the end of a string of summer workdays I would commonly have lost five to seven pounds through dehydration; I gained them back on my days off. I was leaving a brown stain on my bed sheets, so I began showering before bed and washing the sheets every other day. But the stain appeared anyway, and persisted for weeks in the late autumn, even after the rains had settled the dust. Evidently the dust had gone deep into the pores of my skin, and I suspect even now my body contains some of it.

Although my summer habits resembled those of any desert ranger, the foothills of the Sierra Nevada are in no way a desert. Deserts are often defined as regions that receive fewer than ten inches of rain a year, and my part of the American River gets over thirty-four at its lower-elevation, southwestern boundary and over fifty on the higher ground to the northeast. By comparison Seattle averages only thirty-eight. But in Seattle the rain comes year-round, and on the American River almost all of it falls between November and April.

Perched on the 39th parallel, the American River country is neither a southwestern desert nor a northwestern rain forest. Instead it is claimed alternately by season, a sort of Alsace or Lorraine — those European provinces taken by France in the 1300s, won back by the Germans in the Franco-Prussian War, and lost again to France at Versailles in 1919.

In our hot, dry summers and in the sort of dry, scrubby vegetation you see on our south-facing canyon walls, this country belongs to the great Southwest — dusty, parched, and baking, the leaves of its prickly brush and tree species coated with layers of waxy stuff to seal in their moisture. But in the rainy winters and in the lush coniferous forests of our north-facing slopes and shady side canyons, the American River country pledges allegiance to the Pacific Northwest, that nation of Douglas fir, thimbleberry, black bear, salmon, and rain, which stretches from here to southeast Alaska.

By late September, the first thing to change is the wind. Absent for most of the summer, it begins to blow again. One day thin clouds streak the sky, then lower to form a thick, featureless blanket. A little rain falls on a warm night, and when the weather clears the days are still warm but people begin to feel like putting up firewood. By October the nights grow chill and the black oaks on the ridges are tinged with yellow and orange. By November the rains come in earnest. In the woods, the carpet of moss covering rocks and tree trunks that has been brittle and apparently lifeless for months becomes vibrant green again. Bug-eyed orange salamanders and newts make jerky slow-motion patrols across the forest floor. Ferns tremble with drips from the trees. Mushrooms come up. Water falls in diamond ribbons from moss- and fern-covered cliffs and skeins together into creeks, seeking the river. And the roads we rangers travel, which for months have hemorrhaged clouds of soil behind every car, turn to mud.

Steve MacGaff, our supervising ranger, was about forty when I met him, a slight, taciturn northeasterner with a boyish face and a small mustache. He had been left in charge of the American River canyons when the Auburn Dam effort stumbled and the Bureau of Reclamation cut the budget for patrolling the lands it intended to flood. The superintendent, chief ranger, and most of the rest of the staff had been transferred, and by 1983 MacGaff had been left to handle forty-two thousand acres with three rangers, a maintenance man, and a couple of seasonal assistants in the summer. By the time I arrived he'd managed to beg back one more part-time and one full-time ranger, but he spent the rest of his career with responsibilities far greater than men and women of his rank usually had. He was much loved by his subordinates, because he was highly competent at his own work and mostly left us to do what we pleased with ours. In practice we functioned as a sort of consensus paramilitary. The downside of this was that certain things didn't get done at all. But since no one at headquarters had the slightest interest in the land that had been intended to go underwater or us, our omissions went as unnoticed as our accomplishments.

MacGaff was ideally suited for running the place on the Bureau's shoestring. He had a talent for bookkeeping, but more than that, he was by nature a fearsomely parsimonious man. Ranger Ron O'Leary, an avid theorist of the effects of race and national origin on character and personality, said that MacGaff, whom he sometimes referred to as "the Scotsman," was genetically predisposed to his almost pathological thrift by his ancestry. And because no one else volunteered to handle the mountain of paperwork the department required to requisition the smallest thing, MacGaff exercised a virtual stranglehold on our acquisition of supplies and equipment.

Most of the things MacGaff ordered came from State Stores or Prison Industries, and their cheapness was proportional to the disappointment you experienced as soon as you used them. These goods were the kind you might imagine getting under the command economies of communist countries at the time: tires that blew out on our rough roads, truck batteries that didn't have the power to run all of our radios and emergency lights (we were constantly jump-starting each other), steel lockers for our gear that, once loaded, turned from rectangles to parallelograms so their doors would neither open nor close, and the like.

Our raingear was no exception: a two-piece set consisting of a rubbery yellow parka with a hood and a pair of voluminous bib overalls like those used on fishing boats, but of poorer quality. At one of the frequent motor vehicle accidents on our rain-slick canyon roads, they made me immediately identifiable to ambulance attendants, witnesses, and volunteer firefighters from town — as a commercial fisherman who had somehow lost his way to the sea. There was no shoulder insignia on this rain suit, or any place to pin a badge. Worn together, the parka and overalls had the advantage of putting a double layer of protection between the drenching rain and your gun belt, with its pistol, handcuffs, and portable radio. At the same time, should you need any of these things, you would have to undress to get to them, and for a law enforcement officer this constituted something between a constant annoyance and a potentially lethal situation.

These rain suits are among my first recollections of the American River canyons. When I arrived in May of 1986, they hung on a gray Prison Industries coatrack just inside the old kitchen where we dressed each morning. They were covered with orange smears of mud from the floods of February, when the cofferdam had collapsed and the river had very nearly taken Sacramento.

Sacramento was founded in the winter of 1848–49. From the very beginning, the city existed in fundamental denial of the nature of its site. Local Indians knew that the riparian woods around the junction of the American and Sacramento Rivers flooded regularly. No doubt had the city's founders bothered to look, they might have seen little collections of driftwood caught in the lower branches of trees in what is now Sacramento's downtown. But after the discovery of gold up the American's South Fork the previous year, Sacramento's boosters knew spring would bring thousands of gold seekers to California by sea. Arriving at Yerba Buena — now San Francisco — they could be expected to come up the Sacramento River and debark at the mouth of the American, where they'd provision themselves and head upstream to the mines. One of the city's founders had stockpiled supplies to sell at inflated prices, and so that winter he and his associates surveyed the floodplain into town lots for sale, and those who bought the lots set up stores, restaurants, hotels, and livery stables. It was a disastrous choice.

By the time the two meet in the utter flatness of the Central Valley, the Sacramento River, flowing south out of the mountains of Northern California, and the American, flowing southwest out of the Sierra Nevada, are no longer the swift mountain streams they began as. From Sacramento, the Sacramento River must go another sixty miles to reach salt water at San Francisco Bay, and in that distance it loses only *two feet* of elevation at low, summer flows, and only about thirty feet when the water stacks up on itself trying to get out of the valley during floods. To make matters worse, every other river on the west side of the Sierra Nevada and the east side of the Coast Ranges must, like the American, join the Sacramento to pass through the only breach in the mountains around the four-hundred-mile-long bathtub of the Central Valley — an aggregate, before human modifications, of over half of the state's annual rain and snowmelt.

The nature of Sacramento's site was revealed to its citizens in the first winter of the city's existence, when the American and Sacramento Rivers transgressed on the city a mile back from their banks. There were floods again that spring of 1850, in March of 1851, a few days before Christmas in 1852, on New Year's Eve of 1853, and again that March. After a string of deceptively reassuring years in the late 1850s, during which there was much building, in December 1861 and January 1862 the whole middle of the Central Valley became an inland sea sixty miles wide and a hundred long. Scores of people and thousands of cattle drowned, hundreds of homes and businesses were destroyed, property losses equaled a quarter of the assessed valuation of improvements in California at the time, and on Inauguration Day in 1862, the state's new governor, Leland Stanford, made his way to the festivities in a rowboat. Although attenuated by flood control structures, that sort of thing has continued in Sacramento every decade or two until the present day.

And yet the problem with the climate of California was worse than just floods.

During the Gold Rush, with supplies coming expensively by sea around Cape Horn, a market for locally produced food was created and some immigrants took up farming. The new farmers had come from the East, from Europe, and from other places where it rained in the summertime, but in California they had to make it through five or six rainless months every year. Some winters and springs the rains failed to come; fresh on the heels of the catastrophic floods of 1861–62 came the drought of 1864, and the Central Valley was littered with dead and dying cattle and abandoned crops and homesteads. So farmers were soon attracted to irrigation.

By 1854 the first diversion dams were constructed on the American, one of them at the later site of the Auburn Dam. The North Fork Dam was made out of tree trunks stacked in cribs, log-cabin style, and filled with stones. It served ditches and aqueducts that eventually reached some sixty miles in length, and by the late nineteenth century it supplied a growing number of irrigation farms. Never higher than 25 feet, the dam didn't store water, just diverted some of it, and therefore it had no impact on flooding downstream. But by the end of the nineteenth century, with the technologies of concrete and earth-moving machinery, came a solution to both floods and drought: the storage of flood waters behind huge dams for use in the summertime. The advent of electric light and power in the 1890s made the damming of rivers in the mountains triply attractive, and between 1910 and the late 1920s demand for electricity grew at the rate of 10 or 11 percent a year in California. By the mid-twenties surveyors ranged all over the hills, looking for dam sites. One they found was a bowl-like valley around the confluence of the American's North and South Forks near Folsom; another was upstream, in the eight-hundred-foot gorge of the North Fork below Auburn, where the little North Fork Dam already stood.

It took over twenty years for either site to be used. Folsom was the first, and at the time a 340-foot concrete dam was rising from the riverbed there in the early 1950s, its builders believed it would protect Sacramento from floods of a size seen only once in two hundred years, or longer. But before it was even finished, a huge storm over Christmas 1955 filled Folsom's million-acre-foot reservoir in a single week. (An acre-foot is just what it sounds like, the amount of water it would take to cover an acre of land with a foot of water if nothing soaked in, or just over 326,000 gallons. This is roughly the amount an American family uses in a year.) By the following year bills were introduced in Congress to authorize a larger dam upstream at Auburn. It took a while to get Congress's approval, but after Northern California suffered catastrophic flooding again in 1964, by September 1965, Auburn Dam was law.

To construct Auburn's foundations, the Bureau of Reclamation had to dry out the riverbed. So the engineers built a tunnel big to enough to drive a train through, which created a shortcut at a bend in the river through over two thousand feet of canyon wall. When the diversion tunnel was finished, they constructed a temporary earthen dam over two hundred feet high at its entrance to steer the river into it.

Basing their conclusions on only a half-century of data (river gauges had not been installed on the forks of the American until 1911), the Bureau's engineers calculated that the temporary dam — called a "cofferdam" — could safely contain a storm that came only once every thirty-five years on the average, and the main dam could be finished in far less time than that. Most of the time during construction the entire river would flow through the tunnel, but the Bureau knew that every few winters the river's flow would exceed the tunnel's capacity and the water would back up behind the cofferdam and flood the bridge over which ran the only all-weather road to the town of Foresthill, upstream. When the big dam was finished it would inundate the old bridge anyway, so by the fall of 1973 the Bureau finished a new Foresthill Bridge, which soared across the North Fork canyon 730 feet above the river.

With these preliminaries completed by 1975, the Bureau turned its attention to constructing the Auburn Dam itself. But that year the dam ran into major technical problems — these I will go into later — and by February 1986 the cofferdam and diversion tunnel still stood guard over the dam's unfinished foundations, and no water lapped at the tops of the two four-hundred-foot pillars supporting New Foresthill Bridge.

The civil engineers who had studied floods on the American and Sacramento Rivers for over a hundred years before I got there kept beautiful records. I pored over some of them when I began writing this book — nineteenth-century soundings of river bottoms dangerously choked with mining debris from the mountains, meticulous notes on the flood of 1896 made in an engineer's even hand on college-ruled paper, the ribbon-textured typescript of a report on the Central Valley flood of 1907. These and many other records I found neatly stored in document files in agency libraries. There was an awareness of posterity in the care with which they had been conserved.

In contrast, the records of our lives as rangers under the waterline of the Auburn Dam were less carefully kept. They consisted of criminal investigations, accident and coroner's reports, daily patrol logs, correspondence, bookkeeping, attendance reports, dispatch logs, crime scene photographs, tape recordings of interviews with suspects, and piles of manila envelopes and plastic bags containing criminal evidence. When I began going through them in 2001, they were archived in a decaying midden of sagging cardboard boxes, covered with dust and mouse droppings and stacked haphazardly in an unheated warehouse at our ranger station, surrounded by piles of cast-off things for which there could be no conceivable use: bits of long-gone patrol trucks, shotgun racks, pieces of light bars, dial telephones, ancient sirens in tangles of wire. Inside the boxes were more droppings and mouse nests made of our shredded reports. Bundles of each ranger's citations by year scattered like dry leaves when I picked them up; the rubber bands holding them together had disintegrated to sticky crumbs.

Perhaps by the time you read this, they will have been discarded. More likely, they will be decaying in the same location, because a decision to throw them out would be indicative of a culture of housekeeping we never had. Instead, we scattered our effects — papers, bits of trucks, old holsters and radio batteries, locks without keys — in piles behind us in the cobwebby sheds and offices of our compound, living like squatters or transients, day by day, month to month, year to year. I don't suppose it ever occurred to us that later someone would be interested in the grand social science experiment in which we had all participated without knowing it, which answered the question: How do people behave in a condemned landscape?

The boxes from 1986 contained my work, but not until May. However, by virtue of the other rangers' patrol logs and reports, accounts of witnesses, newspaper reports, video taken by television stations, weather maps, river flow records, and my knowledge of the habitual ways in which the rangers rattled in and out of the American River canyons, I am able to reconstruct what happened during the floods of 1986.

In the first days of February it had been raining and snowing hard in the Sierra. On Sunday, February 2, a bolt of lightning hit a tree on the north end of Auburn and from there passed through two houses. In one, it blew off a wall a patch of Sheetrock the size of a door, which flew across the room and struck a thirteen-year-old girl in the face. She was not seriously injured.

On Monday, February 4, a cell of high pressure off the coast provided a break in the weather. Two of the rangers who later became my partners were dispatched into the canyon under the Foresthill Bridge after a report that someone had either committed suicide or jumped from the bridge with a parachute. On arrival, they found nothing. At about nine the next morning one of them set fire to a pile of tree limbs downhill from our office and stood contentedly in the radiant heat, shovel in hand, as he did with each winter's burn pile. That evening, a twenty-three-year-old woman by the name of Karla Jean Eichelberger drove her car to the east side of the Foresthill Bridge. Evidently her fear of heights got the better of her, because just before seven o'clock that evening she shot herself in the head with a .38 caliber revolver. She was found later that night.

On Saturday, February 8, as was their routine on winter weekend mornings, the rangers gathered at the top of the dirt road down to Mammoth Bar. Mammoth was a quarter-mile-long beach on the Middle Fork, about a mile upstream from the Confluence. In the laissez-faire regime of the Bureau prior to the rangers' arrival in 1977, it had become popular with off-road motorcyclists, and the canyon wall above it was now covered with the red gouges of hill climbs, which bled muddy water into the river whenever it rained hard. Under the relentless logic of "It'll all be underwater sometime soon anyway," State Parks could not summon the political will to close it. In fact, over the following years it was expanded. Helpless to defend their ground against this onslaught of off-road vehicles, the rangers exacted increments of revenge in a multitude of small cuts, setting up roadblocks where they stopped every pickup truckload of all-terrain vehicles coming into the area and writing whole books of tickets for offenses such as expired registration and no spark-arresting muffler.

This particular morning, two sheriff's cars came driving down the road. The deputies were looking for three all-terrain vehicles that had just been stolen in town. The rangers volunteered to check the canyon bottom. As so often happened, while looking for one thing, they found another. Down at the river they heard yells for help, and looking across the water, they saw a stranded climber hanging from a cliff on the other side. Dave Finch and his partner drove to the bridge downstream, crossed the river, and made their way up to rescue the stranded man. Later that night one of the sheriff's deputies, a young reserve by the name of Tim Ruggles, was killed when a patrol car driven by his partner skidded off the road and hit an oak tree as they were responding to back up another unit on a theft call.

The next morning the rangers were back at Mammoth Bar, running off-road vehicle license checks on the radio and eating doughnuts and cinnamon twists from Hilda's Pastries in Auburn off napkins on the hoods of their trucks. And again the comfort of their routine was disrupted, when a man named John Carta and several associates towed a trailer onto the middle of the Foresthill Bridge. On the trailer were a specially constructed ramp and a motorcycle. Carta's accomplices set up flares to stop traffic, deployed the ramp, the motorcycle, and two men with video cameras, and situated a getaway car at a trailhead in the canyon bottom. Carta donned a parachute, snugged the harness, mounted the motorcycle, and accelerated up the ramp and over the bridge railing into thin air.

Airborne, Carta pushed away from the motorcycle and pulled the ripcord. His main canopy opened with a crack and a jerk, and he drifted sideways, passing over the live oaks and gray pines as the motorcycle tumbled away from him. It landed on the canyon bottom with a distant metallic crash, a tinkle of flying parts, and a spray of oil and gasoline. The sound drifted up to Carta, mixed with the whisper of the river, as he rode his canopy and a happy wave of adrenaline, tugging the control lines toward a safe landing on the slope below him.

Someone reported the jump to the Sheriff's Department, the sheriff's dispatcher called State Parks, and two rangers were rolled from Mammoth Bar. Arriving at the big bridge at twenty minutes before noon, one of them recorded the identification numbers from the twisted wreck of the 1983 Yamaha at the bottom of the canyon. The other checked the surrounding area for clues and soon found the trailer and its ramp stashed nearby on Lake Clementine Road. The trailer was registered to Carta. Within a few hours one of the cameramen sold his tape to a television station, and there was no doubt who the daredevil was.

That Monday, on the sixteenth floor of the Resources Building, a serpentinite-green monolith in Sacramento housing the headquarters of the State Department of Water Resources, a meteorologist by the name of Bill Mork pulled the morning weather charts off the old wet-process plotter and with growing concern showed them to a fellow forecaster, Curt Schmutte.

There was something jarringly familiar to both men about the pattern, and Schmutte went looking for the old weather maps from the Christmas storm of 1964 — a storm so warm it rained at ten thousand feet above sea level in the Sierra, causing flooding that killed twenty-four people — and the Christmas storm of 1955, which flooded a hundred thousand acres and killed sixty-four. When Schmutte returned, the two men spread the old weather maps out next to the new ones, and after they finished looking at them, Mork picked up the phone and called the National Weather Service's lead forecaster in Redwood City. Comparing notes, the state and federal men agreed: There was something to worry about.

By Tuesday there were high, thin clouds over the American River canyons. The upper-level charts from the National Weather Service showed that a mass of high pressure that had been blocking storms from entering California since February 5 had split into two pieces. A strong westerly flow of warm, moist air off the tropical Pacific had broken through the high in two branches. One branch took a meandering route around the northern remnant of the blocking high through the Gulf of Alaska, where it was chilled, and then flowed south again. The southern branch was charging east toward California through the breech between the two masses of high pressure. Colliding, the two branches formed a deadly pattern, because warm, moist air, when suddenly cooled, can no longer hold its moisture and drops it quickly as rain. The storm would hit the ramp of the Sierra Nevada at an almost perfect right angle, lifting it more quickly than if the storm had struck the mountains at an oblique. The quick push upward into colder regions of the higher atmosphere would increase the storm's violence, as moisture was suddenly wrung out of it over the mountains.

That morning the two rangers who'd written up Carta's jump hiked back up under the Foresthill Bridge to check on his smashed motorcycle. The county sheriff had taken an interest in Carta and planned to remove the motorcycle as evidence. But the rangers found it gone, and when he learned of this later in the day, the sheriff was angry. There had now been at least eleven suicides from the bridge, he told a reporter from the *Auburn Journal,* and all sorts of people were making a hobby out of leaping from it on hang gliders and parachutes. He had no intention of allowing the bridge to become a destination for every crazy person who wished to risk his life, or end it. But, of course, this had already happened.

The storm set in on Wednesday, and Blue Canyon, in the mountains east of Auburn, got three and a half inches of rain. That evening a rosary was recited for Deputy Tim Ruggles, and the rain beat down on the roof of the funeral home where he was laid out, and on the mourner's umbrellas as they arrived. At ten the next morning Sheriff Nunes attended Ruggles's burial. Ruggles had been twenty-three. The grave was filled, and that afternoon the rain beat down upon the fresh mound of earth under a formless sheet of gray sky, and dusk fell early on Auburn. The town got an inch and a third that day.

By Thursday it was raining as if the world was ending. That day a request for a warrant for the arrest of John Carta made its way from the Auburn offices of Sheriff Nunes, across a wet lawn and a parking lot to the offices of District Attorney Jack Shelley, a big, red-faced man in suspenders. By this time the clay soils of the hills around the American River were saturated and every drop was running off into the river. At two-thirty that afternoon enough water was going down the North Fork alone to fill a container the size of a football field with six feet of water in less than a minute. By Friday that figure doubled; by Saturday it tripled.

By Sunday, aircraft over the Pacific Ocean were reporting jet-stream winds of 210 miles per hour at thirty-five thousand feet, and airline passengers from Hawaii were arriving early at mainland airports. Satellite photos showed the normally spiral patterns of clouds coming in off the ocean straightening out into a sort of gun barrel. The gun was loaded with moisture and pointed right at the American River. That day Blue Canyon got over five inches of rain; on Monday it was eight and a half. Now enough water was going down the North Fork to fill that football-field-sized container with a foot of water in less than a second. By afternoon, every route over the Sierra Nevada was closed by weather; huge landslides covered all four lanes of Interstate 80, both lanes of State Route 50, and the main transcontinental railway tracks. For the next couple of days, the park rangers could do little but watch the water rise behind the cofferdam.

By the time of Tim Ruggles's burial, the tunnel around the foundations of the Auburn Dam had reached capacity. As the water backed up behind the cofferdam, the tunnel went under, and big slick logs began racing in crazy circles around an ugly suck hole in the brown water, like the kind that forms over a bathtub drain, but big enough to swallow a cement truck. By Saturday, February 15, one of the rangers recorded on his patrol log that the water behind the cofferdam had reached the Highway 49 Bridge and was rising toward the bridge deck. Just upstream in the Middle Fork canyon a saturated cliff face supporting the Old Foresthill Road gave way, and the road slid several hundred feet down the canyon wall into the river.

Over the weekend no one had emptied the rain gauge at the Forest Service ranger station in Foresthill. By Monday it registered eleven inches of rain. Before dawn that day the deck of the 49 Bridge went under, and at five-thirty that morning Virgil H. Morehouse of Minden, Nevada, became the last person to try — and the first to fail — to cross it.

With his three children in his two-year-old Buick sedan, Morehouse set out across the flooded bridge from the east side, but the vehicle soon stalled. He tried the doors, but the water pressure from outside would not allow him to open them. He tried to roll down the electric windows, but they weren't responding to their switches.

In a growing panic, Morehouse eventually managed to break a window and, towing his children, wallowed to safety along the flooded bridge deck to the Auburn side, where he was assisted ashore by highway patrol officers who had been closing the road. Within a few hours two of the rangers drove back down Highway 49 to find the bridges at the Confluence completely submerged. Morehouse's Buick was nowhere to be seen. When the waters receded on Tuesday, the car was found sitting right side up in the river bottom about 200 feet downstream of the 49 Bridge. Missing from it was Morehouse's Ruger Single-Six .357, a big hog-leg of a single-action cowboy gun that was by far the most popular handgun in the American River canyons at that time.

During the Gold Rush, makeshift towns had appeared along the gravel bars in the bottoms of the American River canyons, providing a range of services to the thousands of men who were excavating the riverbed for gold. By the 1860s most had been washed away at least once by floods, and eventually all were abandoned. The surviving towns of that era are generally located on higher ground, mostly along smaller tributaries. The town of Auburn is one of these. Now the county seat of Placer County, it was established as a mining camp in the spring of 1848 after the discovery of gold in Auburn Ravine Creek. But by 1986 the creek had been paved over to make room for parking, and it now passed ignobly beneath the center of Auburn's historic district in a storm drain.

As the storm gathered strength, the creek got too big for its conduit. A dozen amateur actors were rehearsing a melodrama at the Opera House Dinner Theater in Auburn's Old Town when the creek burst through the back of the building and the stage exploded into the seating area. From there the water picked up tables and chairs and carried them through the front of the building into the street. Outside, one of the chairs was later found to have been propelled with such force that its leg was embedded in the asphalt. Three actors tried to save the troupe's piano by lifting it onto a pile of debris, but the water soon rose to the height of the bar and they swam to safety. Nearby, the owner of the historic Shanghai Bar and Restaurant tried to keep the creek out of his business by barricading the back door. The water soon found its way through the kitchen window, and he fled to higher ground.

The damage in Auburn was limited to the plaza along the bottom of Auburn Ravine, but where the American River flowed into the flat Central Valley, the situation was far more serious. One residential neighborhood near Sacramento State University is fifteen feet below the waterline of a major flood. In the Natomas Basin on the city's northern edges, the figure is twenty. In 1986, only two things stood between Sacramento and that water: a system of levees begun after the flood of 1850 and improved ever since — usually after they failed — and a single dam, Folsom, about twenty miles east of the city.

If much of the energy behind the construction of Folsom Dam had been generated by fear of flooding, by 1986 the dam was operated by the Bureau of Reclamation, an agency chartered to provide irrigation water, not flood control. At that time, of the reservoir's million-acre-foot capacity, only 400,000 acre-feet of space were normally kept available for flood control. But so far that winter had not been a wet one, and the Bureau had begun hoarding water to fulfill its irrigation contracts with farmers. Thus when the storm began, only three quarters of the usual flood storage capacity was available, or 300,000 acre-feet. Further, although the Bureau had its own weather forecasters, the agency based its decisions on how much water to release from the dam on changes in the lake level, not on their predictions. Thus the Bureau's responses were delayed — even to rain that had already fallen in the mountains but hadn't yet reached the lake.

Monday night the inflow to Folsom hit 200,000 cubic feet per second (CFS), but the Bureau could release only 115,000 CFS from the dam, which was the known capacity of the levees downstream, through Sacramento. And so Folsom Lake rose steadily toward the dam crest.

The deathwatch at the Auburn cofferdam began after midnight

Tuesday, February 17. In the predawn hours, engineers from the Bureau set up a video camera along the canyon wall above the dam, intending to learn what they could from its destruction. By five-thirty that morning the sheet of brown water filling the canyon had reached the dam's crest. At dawn the engineers turned on the camera and stood there glumly in the drizzle to witness what would happen. The wind and rain in the canyon bottom had let up, and, looking upstream, the engineers could see a little stream of water trickling over the left side of the dam, reflecting the pale morning sky. The little creek over the dam crest looked peaceful. But the dam was made of earth, and bit by bit the creek excavated a deeper channel for itself, and as it did, its volume increased and the erosion quickened.

It took about three hours for the dam to wash out. In Skyridge, a subdivision of what had been intended to be lake-view homes on the Auburn side of the canyon, people were taking the day off from work to stand on their decks and watch their tax dollars go down the river. Six-packs and bottles of wine showed up. A partylike mood prevailed. After all, this wasn't something you saw every day — a dam failing, one man told a visiting reporter. Beneath them the little creek running down the dam face had become a horseshoe-shaped waterfall that grew steadily, 25, 50, 70, then 100 feet high as portions of the dam collapsed into it. At the bottom the rusty brown water exploded upward in a hellish maelstrom, filling the canyon with an unearthly rumble. There was something strangely beautiful about it — but not for the Bureau engineers. When the erosion finally reached upstream to the lake the dam was holding back, the whole left side of the structure melted in a heroic climax of water and mud, and a hundred thousand acre-feet of stored water roared downstream into Folsom Lake. It was a hundred thousand acre-feet the Bureau hadn't made space for.

William Hammond Hall, one of the great nineteenth-century civil engineers who studied floods in the Central Valley, said there were two kinds of levees: those that had already failed and those that would. Now Sacramento's would be put to the test. When the contents of the cofferdam spilled into Folsom, the Bureau had no choice but to raise releases from Folsom to 130,000 CFS — two and a half times the displacement of an Enterprise-class aircraft carrier every minute, and 15,000 CFS more than the levees were designed to take. Inside the dam a couple of the operators followed a catacomb-like passageway to a door opening onto a steel inspection walkway a couple of hundred feet up the dam's downstream face. Outside, they had to hold on to the railing to keep from being blown off by gusts from the massive pile of whitewater beneath them. The noise and drenching spray were beyond description, one of them told me later, and the whole 340-foot-high dam seemed to vibrate under their feet.

Down the river in Sacramento, five hundred levee-tenders were now deployed twenty-four hours a day to watch for breaks in the tenuous mounds of earth that kept the swollen rivers at bay. On the Sacramento River just upstream of the mouth of the American, they opened a row of escape weirs through the levee, allowing some of the Sacramento's flow to make its way around the city in a sacrificial channel of farmland known as the Yolo Bypass. So much water was coming down the American that now the Sacramento River began flowing backward, from the mouth of the American to the bypass weirs upstream.

Other crews went around reinforcing weak spots as they were reported. As a rule, levees do not fail all at once. Instead, the pressure of the water inside the levee seeks the tiniest breach to get to the low ground outside: a gopher hole, a cavity left by a long-rotten tree root, or a forgotten pipe. Once the water starts moving through, it mobilizes grains of soil and steadily enlarges the hole. Suddenly a little spring appears from the ground, sometimes at a considerable distance from the levees. At first these springs don't look like much, but they're terribly ominous if you understand what they are. At College Park, one of them came up in a resident's front yard. Men in rain suits quickly appeared with truckloads of sandbags and built a wall around it, forming a structure that resembled a small aboveground swimming pool, to increase the hydraulic pressure and slow the flow. The levee held.

With the flood-fighting crews fully deployed, Folsom Reservoir rising toward brimming over, and the water in some of the levees downstream a foot below their crests, there was nothing to do but prepare for the evacuation of large portions of Sacramento. Then, for no particular reason, the storm abated, Folsom's rise leveled off, and in a few hours the rivers began to subside. When it was all over, flood officials remembered for years afterward that the salvation of Sacramento had come not because of something they had done or the strength of protective measures — which were all but used up — but by the grace of God.

The storm of February 1986 deposited half of the American River drainage's average annual rainfall on the basin in eight days. In six days the inflow to Folsom Lake alone totaled over half an average year's runoff from the whole four-hundred-mile-long Central Valley and all of its rivers together — well over three times the reserve capacity the Bureau had been holding in Folsom when the storm began. In a white paper published by the Bureau three months later, the agency celebrated its handling of the crisis and, as a preemptive strike against critics, pooh-poohed the accuracy of weather forecasting as a basis for reservoir operations. But a study by the National Research Council was less congratulatory. On February 4, the study pointed out, the Bureau had been warned in a letter from the Army Corps of Engineers about the danger of encroaching on Folsom's reserve capacity, yet had failed to act. Further, the agency knew for several days before the cofferdam at Auburn failed that it was reaching capacity, yet it had not elected to drain enough water from Folsom to accommodate the cofferdam's 100,000 acre-feet of water.

Regardless of what the Bureau had done to worsen its danger, the storm of February 1986 changed the numbers on flood risk in Sacramento. Folsom Dam and the levees downstream were now given odds of one in sixty-three of failing to protect Sacramento in any given year. The new Federal Emergency Management Agency floodplain maps showed the line of a hundred-year flood reaching out to engulf places that had previously been considered safe. The Army Corps of Engineers announced gloomily that Sacramento was now the most poorly protected against flooding of all major American cities. By May 1986, automobile bumper stickers saying BUILD IT, DAM IT! were everywhere around Auburn, and by March 1987 Congressman Norm Shumway introduced the Auburn Dam Revival Act, federal legislation authorizing the resumption of construction at Auburn Dam. And it was under the shadow of a general certainty among local people that the dam would now be finished that I came to know the canyons it would flood, and the rangers who worked in them.

3 / CAREER DEVELOPMENT

"Hoahh," bell grunted, ambling into our ranger station kitchen one morning in my first month on the American River.

"Hooahh," replied our lieutenant, MacGaff. He was putting on his uniform at a row of gym lockers along one wall.

"Wooahh," I greeted Bell, a little too eagerly perhaps. I was the youngest among them and the new guy.

Bell headed for his locker without further comment. He was tall, dark-haired, and deeply tanned, his inexpressive mouth almost hidden by a drooping desperado mustache — at this moment, any-way. The mustache was part of Bell's ongoing experimentation with facial hair. In the autumn, when he took a few weeks off to go salmon fishing and pheasant hunting, he'd grow a beard. He'd keep it through the winter, maybe until spring turkey hunting, and then shave it off in favor of the mustache again when the heat came. Later, when all his favorite baseball players had a goatee, he grew one. He was highly intelligent and a little shy, and did his best to hide both traits behind a kind of country-boy impassiveness. He played softball after work, had a deadly arm, and was mildly fa-mous among the rangers for a foot chase in which he had thrown his baton — a policeman's club — at a fleeing suspect, landing it perfectly between the man's ankles, which stopped him in his tracks and broke his leg. Baton-throwing aside, Bell hated law enforcement. He was really too nice a guy for it, and empathized almost painfully with everyone he ever had to write up or arrest.

MacGaff was pinning his badge onto his uniform shirt. "Hey, Jordan, I still need a career development plan with personal performance standards from you." Turning to Bell, he added, "And that goes for you too, Doug." He pulled his aged gun belt from his locker and made it fast around his slender hips.

Outside, there was the sound of a compact pickup whining into the graveled yard. A few moments later, Finch came through the door and crossed the worn brown linoleum tile with his peculiar fast shuffle, not lifting his feet. He opened his locker, next to MacGaff's, and began to don his uniform.

Another pickup rattled into the yard. Ron O'Leary appeared, carrying his briefcase. He was in his early forties, but his roundish face and intelligent eyes were set between a neatly combed head of prematurely gray, almost white hair and a well-trimmed beard of the same shade. This, his air of dignified reserve, and the plaid sport shirt he wore tails-out over green jeans and Birkenstock sandals made him look more like a university professor stopping by his office on a day off than a park ranger. In fact, he was the only one among us with a postgraduate degree.

"Woo-ahhh," he greeted the others quietly, twisting the combination padlock on his locker. There were *hooahh*s all around in reply. This standard salutation of American River shift changes lacked the puerile zeal of its military antecedent, as uttered by nineteen-year-old Marines. Here it had a weary, ironic sound, like a man grunting after swallowing something bitter.

"O'Leary's the only one who's turned in his career development plan," MacGaff announced to no one in particular, but good-naturedly.

"Bully for you, Ron," said Bell. "Sissy," he added, with the slightest grin. "What did you put on it, anyway?"

"The usual," answered O'Leary, buttoning up his uniform shirt. "Preserve the status quo."

"Preserve the status quo! Hooahh!" Bell chuckled. He strolled over to a large steel locker just inside the old mess hall adjoining the kitchen, where he pulled a bundle of keys from a clip on his belt, unlocked the door, and removed one of the shotguns lined up inside. "That's what I'm putting on mine," he said, laughing. "Preserve the status quo!" He worked the gun's action, *clack-clack,* to make sure it was empty, grabbed a handful of shells from inside the locker, and fed them expertly into the magazine.

Another pickup crunched into the yard. Sherm Jeffries, from the coal country of Pennsylvania, clomped through the door. If O'Leary didn't exactly look like a ranger, Jeffries, with his flinty eyes, rugged face, neatly trimmed black hair and mustache, and black logger boots, looked every bit the part. He was carrying a reel from a fly-fishing rod, which he set gently on the table. "Check this out, Doug," he said to Bell.

Bell walked over and picked it up.

"Nice," he said, setting it down respectfully.

"Hey, no foolin'," MacGaff continued earnestly through the interruptions. "I need those career development plans and personal performance standards. Bruce's gettin' on my case about it."

"Yeah, yeah," said Bell in mock disdain. He crossed the room with his shotgun slung casually over one shoulder and reached for a set of vehicle keys on a row of hooks by the door.

"Hey!" yelled MacGaff in mock anger as Bell went out the door.

Finch was gathering his equipment. I stood with one boot on the seat of a chair, putting on polish. Outside I heard Bell trying to start one of the old Ramchargers, their paint oxidized in the sun and eighty- or ninety thousand bolt-loosening dirt-road miles on their odometers. The big V-8 turned over reticently for a few revolutions, there were a few expiratory clicks from the solenoid, followed by a final *thhhptt,* and then nothing. I heard the creak of the driver's door opening, then the resonant spring-creak of the hood.

Bam! The hood was slammed violently back down.

"Fuck!" Bell's voice.

Bam! His boot hit the fender.

"I'll jump you, Doug," Finch yelled cheerfully as he shuffled out the office door with his arms full of gear.

As our senior statesman, O'Leary had been designated to show me around. That morning he took me up to Mineral Bar in one of the two newer, more reliable rigs. A production sport utility vehicle must be extensively modified for a ranger's work, and this had been done very well on the older Ramchargers, although now, in advanced age, nothing worked right. The two three-year-old GMCs had shown up just after the budget was cut and most of the staff was laid off and had been given to the senior men, MacGaff and O'Leary. The way they were outfitted was a study in resignation.

Loading my gear into O'Leary's back seat, I noticed there was no metal screen between it and the front seat for safely transporting prisoners. The roof was bare of the usual red and blue emergency lights. There were no spotlights, no alley lights, no flashlight charger inside for night work. The vehicle had no shotgun rack, no baton holders, and no radio scanner. I opened the tailgate to check the rescue equipment. There wasn't any, save for a worn fishing-tackle box containing a few gauze bandages in dusty plastic sandwich bags, a superannuated roll of adhesive tape congealed into a solid lump, some dusty bandage scissors, and a stethoscope. There was a suction bulb for removing meconium from an infant's throat at that once-in-your-career emergency childbirth, but no climbing ropes, river-rescue gear, cervical collars, or splints for the falls and boating accidents rangers see a lot of in mountain canyons.

We left the Auburn office, O'Leary driving — he was an excellent driver — and started east up the interstate toward the mountains. At the town of Colfax we left the highway and followed a single-lane road switchbacking down the wall of the North Fork canyon in the deep shade of a Douglas fir forest. At the bottom we reached the banks of the North Fork. The river coursed by us, clear, fast, and cold, its stony bottom dancing through the prisms of waves. We crossed it on a two-lane bridge. Underneath us dozens of swept-winged swallows dipped and weaved out over the rapids in pursuit of a hatch of dragonflies.

On the far side, we turned left on a dirt track through a line of boulders next to a sign that read MINERAL BAR CAMPGROUND. The campground consisted of seventeen campsites laid out along the track, about half of which were occupied by a collection of faded tents, sagging blue plastic tarpaulins, and wasted-looking old cars, pickup trucks, and vans. I couldn't see a single human being. Somewhere on the other side of a line of alders along the riverbank I heard the hum and rattle of a gold dredge.

"They must be out mining," I said helpfully to O'Leary.

"Uh-huh." He nodded noncommittally. He was a quiet man.

We idled slowly through the campground. Where the road dead-ended I saw two campsites below us, in a dusty basin separated from the river by a windrow of boulders cast up on the riverbank by nineteenth-century miners. In one camp, I saw people.

"Those are the Hallecks, a whole family of them," O'Leary said with no apparent pleasure.

A faded wall tent slumped at one end of the site. From it, a short man, his grubby shirt unbuttoned and his pale belly overhanging the waistline of his grimy cutoffs, made his way toward a short woman in a faded, multicolored muumuu. Her long hair hung forward in greasy strands as she bent over a frying pan on a camp stove. The ground was littered with beer cans. The man was yelling something I couldn't make out, and the woman yelled back, waving a corpulent arm. Continuing toward her, the man threaded his way through four dirty-faced children playing on the ground. They seemed oblivious to the domestic strife. I was reminded — and instantly I was ashamed, for these were *people*, I thought — of looking down into one of those pits at the zoo, where some poor creatures pace out their irritable lives in whatever is the opposite of a state of nature. This was not nature, I thought. This was not a park. My career had hit bottom.

O'Leary got out with his clipboard and bank bag and walked down a little footpath from the road to the campsite. I stood above at the edge of the embankment to watch. Below me, O'Leary greeted the inhabitants with genuine politeness. He waited patiently while they dug around in their filthy duffels and disheveled tent for the camping fees. When they finally pieced together enough change and counted it into his hands, he produced a little receipt, which he filled out and handed to the man. Then he made his way back up the path. When he arrived at the top, he was wiping his hands on his green jeans.

"Even the money was dirty," he said quietly, opening the door of the Jimmy.

We circulated back through the campground one more time, trying to find someone else to shake down. At one site O'Leary left a warning note for nonpayment. At another, a wisp of smoke rose from a scorched milk carton and some potato peelings sizzling on the coals of an abandoned campfire. A pump from a gold dredge lay in pieces next to the fire ring on a grimy tarpaulin.

"Where are they? Mining?" I asked O'Leary. I retrieved my canteen from the Jimmy and poured water on the hot ashes, stirring them with a stick. The fire sputtered and steamed.

"Cheese Day," O'Leary replied, squinting at me through the acrid smoke. He turned to walk back to the truck.

"Cheese Day? What's that?" I asked, following him. We got in.

"That's where they give out government food up in Colfax and Auburn. You know, Department of Agriculture surplus commodities for indigents and welfare miners — generic cheese, dried milk, generic macaroni, sacks of beans."

"Oh?"

"So if you want to collect fees, don't bother coming here on the morning of the first Thursday of the month," he said, putting the Jimmy in drive. We left the canyon, headed back to our office.

Back in the ranger station's kitchen, O'Leary chewed a takeout taco he had picked up on our way back through Auburn and absorbed himself in a paperback Louis L'Amour cowboy novel. I studied him in silence, munching sandwiches I had brought from home. When he finished, he wiped his beard with a paper napkin, put the book away in one of the kitchen cabinets, and walked outside into the covered alleyway between the kitchen and what had been a walk-in cooler, now used for storage. I heard the raspy click of his cigarette lighter. When he finished his cigarette, we got back into the Jimmy and headed for Cherokee Bar.

Cherokee Bar was on the south side of the Middle Fork, about twelve miles upstream of the dam site and four hundred feet beneath the dam's high-water line. It took about forty minutes to drive there from our office. Given a vehicle with decent ground clearance and traction, you could have reached Cherokee Bar from the gilt-domed state capitol in Sacramento in an hour and a half. But somewhere in between, the normative influence of the capitol and its laws was exhausted. In those days the situation at Cherokee Bar resembled those peculiar 1970s Westerns in which the bad guys all looked like armed rock-and-roll musicians.

We crossed the North Fork canyon into El Dorado County, headed east toward Georgetown, and then turned off the main road onto a smaller one, up a gully into the pines. Three and a half miles out the pavement expired. We lurched into a muddy wash surrounded by a thicket of blackberries and Scotch broom and then emerged onto the canyon rim. A meek little state park boundary sign stood to the left of the road, thoroughly ventilated with bullet holes. A thousand feet below, the rapids of the lower Middle Fork glittered in the afternoon sun. From there the road got better, but the improvement was temporary, and three miles farther we rattled down a last precipitous switchback onto Cherokee Bar.

Cherokee was a large sandbar on the inside of a slow bend in the Middle Fork. In front of us the road petered out into multiple sets of vehicle tracks across an expanse of beach, shimmering with heat. In the distance along the water's edge, thickets of willow and a few cottonwoods and alders formed oases of shade. The only other refuge from the withering sunlight was a narrow strip of overhanging live oaks along the canyon wall. In their shade, to our right, stood an outhouse coated with that chocolate-brown paint park maintenance workers use on everything. The outhouse was riddled with bullet holes. On its far side was a campsite: a couple of old pickup trucks parked next to some piles of dredge parts, two wall tents, and a large blue plastic tarpaulin strung between them as an awning. Underneath the latter was a fireplace made of stones, around which were arranged a few threadbare aluminum lawn chairs and a couple of ice chests.

"They owe us," said O'Leary, steering toward the camp.

Before we got closer than seventy-five feet, two dogs, a massive Great Dane and a hulking mongrel, emerged from the campsite and charged us, barking furiously. I looked at O'Leary. He sighed and stopped the truck. The dogs circled us, barking and snarling through the open windows. O'Leary didn't look particularly alarmed. Apparently this was normal.

A man in his early thirties appeared from the campsite and ambled toward us, yelling at the dogs. His long brown hair was tied in a ponytail down his back. He wore a broad-brimmed cowboy hat over a red muscle T-shirt. His jeans were stuffed into cowboy boots. From a cowboy gun belt festooned with bullets around his waist hung a holster containing a huge, long-barreled revolver. The holster was tied to his lower thigh with a leather thong, gunfighter-style.

One fundamental of all parks, state and national, is that they exist to preserve wildlife. Further, parks are used intensively by hikers, horseback riders, mountain bikers, and boaters, and this kind of recreation is generally inconsistent with gunfire. Most park system regulations are designed to promote civility between users and make casual visitors feel at ease. So, not surprisingly, it's illegal to walk around wearing a pistol at almost every park in the United States — outside of Alaska, where subsistence hunting is common and some people feel the need to have a ready defense against grizzly bear attack.

When an armed man approaches you on foot, you don't stay in

your patrol car. Perhaps counterintuitively, two armed men in a stationary motor vehicle are no match for a single man on foot. Wedged between your seat, dashboard, steering wheel, and doors, you make a fatally predictable target. It's surprisingly difficult to draw your weapon from the type of holsters we use when seated in a vehicle. If you do manage to get your gun out, seated facing forward, your field of fire is extremely limited. You are entirely visible to your opponent through the glass, but fire a single round from your own gun through one of those windows and immediately you will be deafened, and your aim may be warped by the bullet's impact with the glass. And the thin steel of an automobile body affords little actual protection from modern firearms fired from outside. You had better get out.

I watched O'Leary for a signal. He sat quietly with both hands on the steering wheel, watching the man with vague interest as the latter approached his open window. The Great Dane and the huge mongrel raced back and forth from one side to the other, pawing at the doors. Arriving at O'Leary's window, the miner grabbed at the Dane, and I saw his shoulder jerk as the animal strained against his grasp on its collar. It was still barking furiously.

"Buck! Shut the fuck up!" the miner yelled down at the dog.

"How you doing today?" O'Leary said casually to him. The dog whimpered. The other one was over at my window, showing me its teeth.

"Okay, I guess," the miner replied to O'Leary. "What's goin' on?" Now both dogs were barking again.

"Buck! Moocher! No!" the miner yelled, swatting at the Dane below O'Leary's window.

O'Leary reached for a clipboard between us on which he recorded who had paid and who hadn't. He showed it to the miner. "Seems like you haven't given us any camp fees for a while."

"Yeah," replied the miner. "I was gonna take care of that, but my check hasn't come yet. Scott went up to Georgetown to get our mail. Maybe you can come back."

As this exchange took place I was becoming increasingly uncomfortable. I couldn't see the gun or the miner's right hand. It just didn't seem dignified to offer ourselves up for slaughter like this. Even if things remained friendly and the gun never came out, we were at a conversational disadvantage, because everyone knew we were in no tactical position to make any demands.

"Well, let's get this taken care of or you'll have to leave pretty soon," O'Leary was saying to the miner. He reached for the gearshift on the column. The miner nodded and stepped back, and O'Leary turned the Jimmy toward the river. The gunman watched us for a moment, then began trudging back to his campsite. Turned loose, Buck made one more charge, barking at our rear bumper, then bounded after his master.

Down along the riverbank, we found another camp. Nearby, in the river, two men were running a suction dredge. The dredge was like a small barge on two metal pontoons about ten feet in length. On its deck were mounted a gasoline engine driving a powerful water pump and a sloping aluminum trough. One wet-suited miner was working face-down in the water as we drove up, his mask supplied with air by a small hose from another pump on the dredge's deck. He was maneuvering a much larger, ribbed hose like a giant vacuum cleaner along the river bottom, sucking up rocks and sand. The other man stood on deck watching the material the first miner sucked up rattle down the trough and back into the water, where it formed a muddy plume downstream. Apparently the gold had a way of settling out behind ridges in that trough. The motor and the rattle of the stones made an unholy din. When the miner on deck saw us coming, he yanked on one of the hoses, and when the other surfaced he shut the machine off. The two men waded ashore to meet us. I noted with satisfaction that they were armed with only diver's knives.

O'Leary got out of the Jimmy and collected some damp dollar bills. Then he asked to see their dredge permit. One of the men produced a plastic bag containing some dog-eared papers. O'Leary removed the papers, looked at them cursorily, then handed them to me: a form filled out in blue ink, bearing the imprint of the California Department of Fish and Game, stapled to a few pages of regulations. I gave them back to O'Leary, who gave them to the miners. We thanked them and left.

I looked over at O'Leary as we drove up the beach. "Ron, why does Fish and Game — part of our own Resource Agency — issue permits to dredge for gold here if it's illegal to use even a hand tool like a shovel to disturb a riverbank under our own laws?"

O'Leary sighed. "Welcome to Auburn" was all he said.

We approached the bottom of the road up the canyon wall.

I looked over at him again. "Aren't we going to do anything about the gun?"

"You can if you want to," he answered.

"Okay then, I will," I said.

O'Leary glanced at me through his aviator sunglasses — questioningly, I thought — then turned the Jimmy toward the first camp. I called in the contact on the radio, then flipped the switch to public address. "Mr. Fowler," I announced through the speaker on our front bumper, reading the man's name on O'Leary's campsite registration sheet. "Could you please tie your dogs up in camp and come meet us here when you're finished?"

The miner appeared from behind one of the tents, still armed. He shrugged his shoulders and lifted both palms, questioningly.

"Just tie them up, please," I repeated through the PA.

The miner called the dogs and busied himself rummaging for ropes and making them fast between the dogs' collars and the trailer hitch on one of his trucks. When he finished, he crossed the sand toward us with a pained expression. I got out and faced him from the other side of our vehicle's hood, where the engine block afforded some cover if I needed it. He looked at me warily.

"Is there a problem?" he asked, arriving at the Jimmy.

I stepped toward him to stand directly to his right, within reach of his gun hand.

"Could you put that gun on the hood for me?"

"Why? Is there a problem?" His face darkened.

"You've got a gun."

His eyes narrowed. "So? I haven't done anything illegal."

"You can't carry it in this or any other state park campground," I replied.

"This is a state park? No way!" said the miner. "Where's the sign? I thought this was just the American River. Anyway, it's never been a problem before — like when you were just here a few minutes ago."

He had a point, but you've got to start somewhere, I thought. Beads of sweat were breaking out on my face. O'Leary was out of the Jimmy now, but said nothing.

"Look, just put the gun on the hood, then we'll talk," I told him.

He gave me a don't-you-ever-turn-your-back-on-me look, then slowly unholstered the gun and clunked it down on the hot green steel of the truck's hood. I moved for it, swiftly but with studied nonchalance. Once it was in my grasp, I spun the cylinder and dumped the ammunition. It was loaded with six .357 hollow-points.

"Could I get some ID?" I asked him.

He gave me a defiant look as he handed me his driver's license. I called in a warrants check on him and ran a stolen-gun check. Miner and gun were clear, so I began filling out a citation. When I finished, I turned my citation book toward him.

"What the hell is this for?" he asked. "I haven't done anything wrong, and I *told* you I'd pay you next time you come. You don't have to be a prick about it."

"Your dogs are off the leash. You're carrying a gun. It's a state park."

"The hell it is — I know better than that. It's a dam site." The man looked at O'Leary and his tone changed. "Hey, Ron — you know us. Tell your rookie to leave me alone."

"Just sign the ticket. It's not my call," O'Leary growled through his beard. It was plain he wasn't enjoying himself. I couldn't tell if he was more irritated at the miner, or at me for disturbing his peace.

Turning back to the miner, I continued. "You have been cited into Georgetown court; your appearance date is —"

"You can *bet* I'll be there," he snapped back. He snatched the citation book and the pen I proffered out of my hands. "See you in court, asshole!"

He scrawled his name across the bottom of the ticket and thrust the book at me, poking me in the chest with it. I tore off his copy and handed it to him. He wadded it into his pocket. "Now give me back my gun," he demanded.

"Well, that's not exactly what's happening today," I replied. "I've taken your gun as evidence. If you have no prior felonies, you can ask the judge to release it back to you. Then we'll give you back the gun."

The miner gave me a withering stare, then turned and stalked back toward his campsite. The dogs barked happily. We got back in the Jimmy. I put the gun in the back seat. We pitched and rattled back up Sliger Mine Road. For the rest of the drive to the ranger station O'Leary didn't say much.

Back at our compound I found Steve MacGaff working at his desk. I told him I'd seized a gun, and I asked him where the evidence lockers were and who the evidence custodian was.

"Come with me," he said, getting up.

I followed him back out the door and up to the mess hall. He led me into the back room to a row of gray gym lockers and unlocked a cheap padlock on one of them. Inside were three things: a chrome-plated semiautomatic pistol, an empty beer can around which was wrapped a rubber band holding a scrap of paper with a citation number scrawled on it, and a crumpled paper lunch bag containing several pop-bottle rockets.

"You can put it in here," he said. "And you can be our evidence custodian."

Maybe a ranger career is in trouble as soon as you have one.

In the beginning you just have a job. You work seasonally. Most of the rangers you see in national or state parks are seasonals — in national parks, even a lot of the rangers with guns are. These men and women have put themselves through law enforcement academies and emergency medical technician courses at their own expense and on their own time. Seasonals jump out of helicopters into forest fires and rappel down cliffs to save stranded climbers. Most of them have no retirement plans or medical benefits. But the good ones are so highly sought after that they can have their pick of the most glorious forms of employment on earth: they rock climb for the government in the Grand Tetons and Yosemite Valley, scuba dive on government time in the Everglades and Channel Islands, and go whitewater rafting and kayaking in the Grand Canyon on government per diem.

They — we, for I was one of them before I went to the American River — keep their packs, climbing gear, sleeping bags, wetsuits, and milk crates full of dehydrated food in a van or a pickup with a cargo shell at the park's migrant-worker housing. In the summer they work as wilderness rangers or fight fires in the mountains. In the winter they migrate to the desert or the beaches, or work ski patrol and avalanche control at ski areas.

But after a few years of this, you realize you aren't getting anywhere and have no job security, and you start looking for a permanent job. Before, as a senior seasonal, you could have your choice of assignments. Now, as a junior permanent, you're back at the bottom of the heap. So you gladly take what you can get. Then, to get back to the kind of places that were the whole point of rangering in the first place, you begin to make calculated moves instead of moves of the heart. That's when the trouble begins.

Edward Abbey, writing of his experiences at Arches National Monument in the early 1960s in that greatest of all ranger books, *Desert Solitaire*, eschewed permanent employment. The poet Gary Snyder, who at once formed and echoed the environmental ethic of a generation, worked for the National Park Service on seasonal trail crews in the Sierra and as a seasonal fire lookout in the North Cascades. Under Snyder's influence, Jack Kerouac later took a Cascade lookout job too. Neither man probably ever considered going permanent. As I write this, Snyder is an old man and Kerouac is long dead, but you can still see their North Cascades from Sourdough Mountain. Abbey is dead, but the Arches National Monument he loved lives on. By 1986, however, Ron O'Leary, Steve MacGaff, Doug Bell, Sherm Jeffries, Dave Finch, and I were all permanent rangers on a temporary river. For most of us, our career prospects ended when we went there. The Auburn Dam site wasn't the kind of place that looked good on a resume. The department preferred to think of its rangers chatting with families in neat little picnic grounds or giving wildflower walks. Most of us were never promoted again. What we did there mattered only to us, and to the river.

I never saw the career development plan Bell gave to MacGaff. But one day the following winter I heard the table saw running up in the metal-roofed workshop at the top of our compound. Stepping in through the big sliding doors out of the rain, I saw a fire crackling in the oil-drum woodstove. In the middle of the room Bell was cutting pieces of plywood. Some were already assembled nearby. He was making birdhouses, but they were much larger than normal. I asked him what he was up to.

"Duck houses," he said, "for my career development plan."

A few months later I was walking the shoreline of one of the ponds on the headwaters of Knickerbocker Creek when I came upon several of Bell's boxes nailed to pine trees. By then I knew what they were: nest boxes to encourage wood ducks — the most exquisitely gaudy waterfowl ever seen in the American River country, with their lemon yellow, cinnamon, white, and shimmering green and blue feathers and multicolored cloisonné bills — to settle and raise their young at our ponds. I didn't get too close for fear of disturbing the occupants, but later when I asked him, Bell told me that a couple of duck families had accepted his invitation.

In a way I wish that had been my project, for it was all about birds instead of laws and lawbreakers. Bell also ran the boat-in campground at Lake Clementine, and he later became a prolific builder of hiking and riding trails. The trails and the nest boxes were Bell's signature: Faced with all the malevolence of humankind and their dam, he still genuinely liked people, but he much preferred to deal directly with the land and its creatures.

I never saw O'Leary's plan either. But when I asked him for advice on what to put on mine, he recited the following phrase from memory: "I will meet or exceed the standards of performance for my class." Later, when I asked Finch about that phrase, he smiled and said, "That means 'maintain the status quo.'"

And that O'Leary did admirably. Day in and day out for decades he went to Mineral Bar, Yankee Jims, Ruck-a-Chucky, Upper Lake Clementine, and Cherokee Bar, where he patiently collected crumpled dollar bills and unfulfilled promises from the drifters and miners with the greatest courtesy. He did all of the accounting for camp fees and made up the bank deposits, and when once he was audited, not a penny was missing. He was our pistol instructor and a really competent shooter. For some reason he was fated always to be the one who picked up the paper on those Saturday afternoon fights between drunken young men on the river beaches. These were the most hopeless sorts of investigations, for there were endless numbers of involved parties and witnesses whose slurred statements all had to be written down. O'Leary knew that the district attorney would never prosecute anyone, for there were no clear perpetrators or victims in these things, just winners and losers cut from the same belligerent cloth. Yet he was as diligent in reporting on them as he was in his other work. He was respected both by citizens and by fellow officers, and senior cops from other departments always greeted him first when we worked a scene together, for he was possessed of that amiable reticence seasoned cops admire in others.

But what he had been through in those canyons by the time I met him had taught O'Leary that the government cared not a stitch for this river — he was right about that — and he had decided not to get himself killed for it. And so he maintained a compassionate truce with the lost and hopeless, the minor misdemeanant and the drunk, and unless forced by circumstances beyond his control, he didn't go out of his way to pry into their activities, providing they behaved themselves and paid their camping fees once in a while.

But I was young and brash and I had no real understanding of what O'Leary and the others had been through before my arrival. I thought that what was now forming in my mind was original — the idea of making a real park out of this hopeless dam site, whipping it into shape with diligent enforcement of park regulations and driving off the scary people so that birdwatchers, kayakers, and families would feel safe coming to it. Now, of course, I know all of the rangers had started with that notion.

And so I set out to violate O'Leary's truce. Because I hated begging camp fees from armed men, I began systematically seeking out and seizing every pistol, rifle, shotgun, knife, dagger, brass knuckles, or club I could get my hands on. In the next ten years I confiscated 125 weapons in criminal cases. To store them, I put up evidence cabinets and, eventually, a property room.

Because I didn't want to be seated in a truck while my partner negotiated with an armed man for camp fees, I put down on my career development plan that I wanted to become a defensive tactics instructor, and that winter I went off to school for training. For the next several years I stood over my fellow rangers in long practice sessions on wrestling mats until their wrists and knees ached, drilling them on how to take away a gun from an armed man, what to do if he tried to grab their gun, and how to search and handcuff him safely. They must have thought I was a prick. I guess I was. But they were good-natured about it and trained hard, and we began to look like we knew what we were doing. Meanwhile, MacGaff worked his accounting magic on the Bureau to replace two of the old Ramchargers. When the new Jeeps came in, I found a welder who built racing cars down in Sacramento, and we put together the first of a whole string of very businesslike new patrol rigs, with all the trimmings.

For the rest of that summer and the next I went back to Cherokee Bar alone, over and over. With a shotgun in my arms and my heart in my throat I tiptoed down the riverbank through the brush, seeking out the miners' illegal camps and dredging operations. I arrested them on their various warrants — a lot of them seemed to have had previous problems with the law. I towed their vehicles away.

One day it was Sherm Jeffries who went to Cherokee Bar to collect camp fees, and when he arrived at that troublesome campsite, he got bitten by the Great Dane. When he recovered, he and I went back there together. The dogs came out to meet us as usual, followed this time by several miners. We faced each other and words of unhappiness were exchanged. Between us, the Dane started growling and lunged at my crotch. Without thinking, I hit it hard on the nose with my aluminum ticket book. The animal drew back, whimpering. I wrote the owner another ticket for dog-off-leash. Within a week he filed a grievance with Internal Affairs. It was eventually dismissed. After that my life was threatened several times. One of the Cherokee Bar miners told me he'd been a sniper in Vietnam and one day soon he was going to do what he did best — to me. But in the end, Cherokee Bar began to turn around.

In my first eight months in the beautiful desolation of the once and future Auburn Reservoir, I made twenty-nine arrests. Bell made one, O'Leary three, MacGaff six, Finch — who was spending much of his time at the union office down in Sacramento working for the betterment of all rangers — ten.

At least fifteen cars had their windows broken and their contents stolen that year in our canyons. I suspect there were many more, but the reporting rate was miserable. There were eight assaults and batteries, two rapes, a grand theft, a wife beaten by a husband, one arson, ten drunk-in-public arrests, twelve cases of vandalism — mostly shot-up signs. The operators of ten motor vehicles were arrested for drunk driving. We apprehended fourteen people on arrest warrants for offenses they'd committed elsewhere; of those, twelve were misdemeanors, two felonies. We recovered nine stolen cars and investigated the deaths of four people in accidents or suicides. Twice that year we were involved in high-speed car chases.

We know these things because of what Finch put down on his career development plan. His project involved entering selected data about every incident we rangers handled into a computer database program. From 1986 on he distributed an annual digest of statistics on how people had behaved on the condemned ground of the Auburn Dam site — and the news wasn't good. Of course, this was before Microsoft Access and Filemaker became the industry standards in database programs, and the name of the software into which Finch entered every event reported by the permanent rangers on that temporary river was one most people wouldn't recognize today.

It was a program called Paradox.

AT SOME POINT in the last five million years as it flowed southwest out of the Sierra Nevada, the American River fell in with a line of secret cracks in the earth called the Gillis Hill Fault. A river always takes the easiest way to where it is going and, like some people, will exploit any weakness in its surroundings to get there. So just east of what is today the town of Colfax, the North Fork insinuated itself into the promising cleft of the fault and began running along it. But as it cut downward the river seems to have slid sideways, for the rocks along the fault dip east, at a right angle to its channel. Eventually the North Fork no longer ran along the fault but ran parallel to it. And as is also true of people, rivers sometimes keep doing things long after the reasons for doing them are gone. So for many years — hundreds of thousands, perhaps even a million or more — the river has continued to flow that way.

From the time of the Gold Rush, the five miles of the North Fork along the Gillis Hill Fault were known to be rich in gold, and that reputation persisted even after the North Fork canyon was to be inundated by the Auburn Dam. By the time of my first patrol of it in June 1986, that part of the river had long been one of my fellow rangers' more troublesome responsibilities. True to form, during that first reconnaissance I caught sight of a squatter's shack along the canyon wall and a gold dredge below it in the river. Back at the ranger station I found no permit for a mining camp there, so in July, Doug Bell and I hiked back up the river to investigate. And in this simple way began a dark chain of events that continued into the fall of that year.

To hike to that part of the North Fork, Bell and I would start from the bridge where the road between Colfax and the ghost town of Yankee Jims crossed the river. Completed in 1884, the Colfax–Yankee Jims Road consisted of a crooked shelf barely wide enough for one car, chipped into the slatey shale wall of a tributary called Bunch Canyon with the typical economy of roads built with hand tools and intended for use only by horse-drawn vehicles. Leaving the outskirts of Colfax on it, you passed a couple of abandoned shacks and a collection of junked cars riddled with bullet holes. Beyond these you came to an occupied cabin separated from the road by a tall board fence, on which the resident had painted in large white letters: HAVE A FUCKED DAY!, TRESPASSERS WILL BE EXECUTED, and I SHOOT FIRST AND ASK QUESTIONS LATER. Beyond that outpost you passed several abandoned mine shafts and our shot-up boundary sign, and eventually you reached the bottom of the canyon and Yankee Jims Bridge. Completed in 1930, the bridge's creaky deck depended from two rusty cables about seventy-five feet above the rocky breach full of rushing water, and however tentatively you drove onto it, the whole structure bucked and swayed underneath you in the most unsettling way. In all, Yankee Jims was a forbidding place for a ranger in those days.

On the far side of the bridge that morning, I took a 12-gauge shotgun from our truck and slung it over one shoulder. Bell carried only his revolver. Leaving the road on foot, we forded Shirttail Creek, which tumbled into the North Fork from a rocky slot just upstream of the bridge. From there we followed an old miners' trail up the canyon wall, past a series of ledgelike indentations supported by rock retaining walls — the footings of Gold Rush miners' cabins.

As we started up the canyon, it still mystified me why we tolerated mining. Any such activity was clearly in violation of not one but several state park laws. Since my first look at suction dredging, I had questioned my fellow rangers about it several times. I had received only the most cryptic responses. But the mere mention of mining seemed to incite in them such a mixture of smoldering anger and hopeless resignation that I soon abandoned the subject. It wasn't until a decade and a half later — in the course of research for this book — that I finally learned what had happened to those men.

In 1965, when the legislation authorizing the Auburn Dam was signed into law by President Lyndon Johnson, the price of gold was just over thirty-five dollars an ounce. A decade passed, and the project ran into extended delays in construction. Most landowners in the canyons upstream had been bought out or condemned by then. Meanwhile, by 1976, gold prices quadrupled.

Although large-scale gold extraction was essentially a dead industry in California, unemployed men, men running from child support or the law, and those who were simply square pegs in the round holes of society had never stopped coming to Northern California to prospect. By the 1970s the availability of portable suction dredges and scuba-diving gear made it easier for small operators to recover gold from rivers, and with no one around to tell them to leave, miners equipped with this kind of gear settled in the site of the Auburn Reservoir as squatters. Semi-permanent camps and shacks appeared all over the place. Untended campfires ignited wildfires. Piles of garbage and old cars were dumped down the canyon walls. Guns were pulled in mining-claim disputes. Murder victims began turning up in shallow graves. The Bureau of Reclamation didn't have its own rangers, so as the North and Middle Forks descended into anarchy, the agency tried bringing in federal marshals for occasional sweeps, but that didn't even put a dent in the problem. What was needed was a constant law enforcement presence.

Bureau reservoirs were usually handed over to federal or state park agencies to operate as recreational lakes upon their completion, and at Auburn the agency chosen for that job in a decade-old contract had been California State Parks. So in 1976 the Bureau re-negotiated the contract with the state to bring rangers in earlier. Between 1977 and 1982 a park superintendent, a chief ranger, two supervising rangers, seven police-trained patrol rangers, a clerk dispatcher, a couple of maintenance men, and several seasonal helpers were sent into these canyons. To bring the area under their authority, in 1979 the place was designated a State Recreation Area.

What ensued was undoubtedly one of the strangest stories in the history of the century-old park movement. The laws and authorities given to state park rangers in California had been crafted to preserve cherished landscapes in perpetuity, but they now applied to a pair of river canyons the government intended to put underwater. Not surprisingly, the rangers found it difficult to convince violators, the judges who heard their criminal cases, indeed even their own upper management, why preservationist anti-mining laws should be enforced there. To complete the rangers' misery, by the early 1980s, the new Reagan administration and Congress began drastic cuts to domestic spending. In 1982 the Bureau slashed the budget for Auburn State Recreation Area — federally funded by contract — and the superintendent, chief ranger, several rangers, and administrative staff were transferred out. At the same time the price of gold had really skyrocketed, peaking at $850 an ounce — twenty-five times what it had been when the dam was begun. The social and environmental effects of the resulting gold rush overwhelmed the remaining rangers, and in the winter of 1982–83, seeing no way to fulfill their legal obligations, they closed the river indefinitely to mining and camping.

The hue and cry that followed was deafening. State Parks' Sacramento headquarters was soon deluged with petitions carrying over three thousand miners' signatures, asking for a reversal. At a hearing before the State Park Commission in 1983, miners aired their grievances. Among them was the assertion that State Parks, not gold dredgers, was responsible for polluting the American River — by failing to provide outhouses and forcing miners to use the riverbanks for a latrine.

Whatever the merits of this and the miners' other arguments, State Parks was not known for political heroism. The agency's director immediately rescinded the closure, proclaiming that the miners' camps and dredges could remain on the river while under study — a study that is apparently unfinished over twenty years later, because results have never been presented. (Some of the dredges are still there.) By 1984 the only curb on mining was a rule that no one could camp more than thirty days at a stretch in any state park, which in practice made it hard to set up and run larger mining operations. Still, you had to *find* the miners' camps to enforce it. To further placate the miners, State Parks' director appointed a sort of outhouse czar from headquarters, and under his direction Auburn State Recreation Area — which otherwise lacked the most rudimentary facilities — received a diaspora of privies. Some were installed in places so remote they had to be flown in by helicopter; unused, they were soon overgrown with wild grape and blackberry vines. Some were installed too close to the river and were carried away by floods; for several years the remains of one could be seen perched in a tree near Lake Clementine. Some were used for target practice. Others disappeared entirely, and no one knew exactly what had happened to them.

The canyon was a thousand feet deep, and so narrow at the bottom I felt as if I could reach across the river and touch the opposite wall. Thirty feet below us, the North Fork sluiced through its tight channel of water-polished blue-gray bedrock. The noonday sun seared through the breaks in the trees. Sweat trickled down the inside of my bulletproof vest and made a dark line down Bell's uniform shirt — he preferred to go without the insufferably hot body armor.

A mile up, our path degenerated into multiple trails meandering through a steep forest of live oak and buckeye. We forded another creek and eventually came to yet another fork in the trail. Cut branches had been piled on one side to discourage casual exploration. We pushed them aside and made our way up that side of the fork. The shack was so well concealed that we didn't see it until we were twenty feet away. There was no flat ground in this canyon, and like those of Gold Rush miners, this shelter had been constructed on a bench cut into the canyon wall, supported on the outside by a rock retaining wall. We approached cautiously and I rapped on the plywood door. There was no answer. We let ourselves in.

The shelter's frame was fashioned from peeled fir poles and a few incongruously expensive redwood four-by-fours, and it was sheathed in clear plastic and small pieces of good plywood that seemed to have been salvaged from another building. A roof consisting of a canvas tarpaulin stretched over a row of sapling poles sloped back to meet the cut in the canyon wall that formed the interior wall on the uphill side. The interior was heated by a woodstove made out of an oil drum and furnished with a kitchen counter, two bed platforms, and a couple of rude tables.

Bell was peering at the four-by-fours and at the odd-sized pieces of plywood with a curious expression. He walked back outside and I followed. At the far end of the ledge the miners had dug a pit toilet, a hole in the earth covered with a sheet of plywood, at the center of which stood a semi-conical fiberglass outhouse pedestal with a white enameled-steel toilet seat.

"Hey — this place is made out of our outhouses!" Bell chuckled.

"No!"

"No question about it. Look!" he said, pointing to the nice redwood four-by-fours, the plywood, and the fiberglass outhouse pedestal. All state issue.

Back inside, we inspected the shack's contents. On each of the bed platforms was a grubby foam mattress over which was laid a sleeping bag. Lifting one, I discovered a sheathed machete, and under the other, a loaded 16-gauge shotgun. I unloaded it. Nearby I found a box of .30-.30 rifle shells, but no rifle. Two deerskins with the hair on — no doubt poached with the .30-.30, an ideal deer gun in such country — were rolled up under one of the beds.

Bell was going through a pile of mail and papers.

"Look at this," he exclaimed with a slight grin, handing me a sheaf of dog-eared yellow forms. "They left us their jail booking sheets so we can ID them."

"How considerate," I answered. I accepted the papers and began leafing through them.

The first booking sheet recorded the arrest of one Jerry Ralph Prentice, age thirty, on charges of public intoxication and disturbing the peace at a Colfax bar called the Station House. There was another booking sheet, for a man named Richard Samuel Marks, on similar charges. I went over to the table where Bell had found the papers and examined the rest of a pile of items on it. They included a bottle of pills from a Colfax pharmacy — tetracycline, an antibiotic; the patient's name was Richard Marks. Later, I would frequently find antibiotics in miners' camps. The cuts and scrapes and the ear infections they got from working long hours underwater didn't heal well in the damp canyon bottoms where they lived.

Underneath the pill bottle was a small ledger book. I opened it. It was a diary. Leafing through it, I read the most recent entries aloud to Bell as he continued looking around the shack.

". . . Jerry and I got seven pennyweight in dust and flakes and a nice little nugget today. Went to town and got drunk. Bought some stuff. Saw Kenny and a girl."

The last entry was dated July 24, four days before: "Partied at the Stationhouse Saloon. Don't know why, but Charlie got mad and knocked me down. Might have broken my shoulder."

"That may explain their absence," observed Bell.

When we had finished going through the miners' meager belongings, we closed the door. I fixed a warning notice to the outside for multiple violations of park law. I wasn't planning on getting shot with that shotgun next time I visited, so I seized it as evidence, leaving a receipt. Then Bell and I hiked back down the canyon, maintaining a wary silence in case we should meet the miners and their .30-.30. coming back in. But the hike passed uneventfully.

———

For another month we rangers were kept busy with what I gathered were the usual summer disturbances: loud parties at midnight in the campgrounds, car wrecks, random gunfire, petty theft, the occasional swimmer swept away in the river. The wooly sunflowers that had covered the road banks in May and June were gone now, as were the *Brodiaea* lilies that had gone off like violet fireworks above our meadows just as the grass turned from green to brown. By August the meadows bleached pale blond in the overwhelming brightness of the summer sun. Nashville warblers, Pacific Slope flycatchers, Bewick's wrens, black-headed grosbeaks, and the other birds whose exuberance filled the forests of early summer raised their fledglings and flew away.

After Labor Day the park settled down. At ten o'clock in the evening on the Thursday after the holiday, Finch and I were bumping down Yankee Jims Road on patrol. Our headlights swept the dark cliffs, and the cloud of dust behind us was lit by the lurid glow of our brake lights. Just inside our boundary I saw something sparkle in the woods below the road. We stopped and got out to have a look.

From the edge of the road, our flashlight beams fell on a beige Chevrolet compact upside down below us, leaning against a tree. As we picked our way down the embankment, I steeled myself for what we might find inside. But there were no bodies or splashes of blood, just crumbs of safety glass, a scattering of personal items, and a woman's handbag. We climbed back up to the road and went through the handbag. Inside was a clutch purse containing the driver's license of a Mary Elaine Murphy of Colfax. I radioed Roberta at the dispatch office and asked her to check on that name and license and send a tow truck. When the wrecker arrived we had the car winched up the bank and taken to an impound yard for investigation. Then we drove on to the river to look for witnesses.

It was close to midnight when we got to the bridge. The river sounded louder than usual in the darkness below us. On the far side we found an older pickup with a camper on the back, parked along the shoulder. We lit it up with our spotlights and rapped on the quilted aluminum door of the camper. The door opened and a blond man in his late twenties squinted into the glare. I recognized him from another, unrelated incident. He mumbled that sure, he knew all about the wreck up the road. A miner named Ricky Marks had driven the car off the cliff, he had heard. I said the name sounded familiar.

"You oughta know him, Smith — you took his shotgun," the tousled man said, rubbing one crusty eye. "Anyway, like I was saying, Ricky was drunk, and after the wreck he stayed with me and Kenny down here by the bridge insteada going back up to his cabin. Middle of the night there was a bunch of noise and we all woke up. Ricky was lying in his sleeping bag, surrounded by men with guns — serious guns. One of them grabbed him by his hair and shined a flashlight in his face, holding a gun to his forehead. They had some woman with them. 'Is this him? Is this him?' the one who had ahold of Ricky kept yelling. She didn't answer. They kept yelling at her and finally she said yes, it was him. Then they kicked the shit out of Ricky, and when they were finished they told us not to say anything about it or they'd come back and kill us. Then they all got in their trucks and left. I couldn't sleep for the rest of the night. Ricky was lying there bleeding and crying. He didn't want to go to the hospital. Anyway, what was I supposed to do? They had guns."

"Is he dead?" Finch asked matter-of-factly.

"Nah. But he was one sore miner in the morning. He looked horrible, face all fucked up," the blond man answered.

We got a name and the usual general-delivery miner's address from our witness, thanked him, and started back up the road. On the way home Roberta called on the radio. Using the address on Mary Murphy's driver's license, she had found the woman's ex-husband in the directory and phoned him. Evidently he hadn't seen Mary for months. To the best of his knowledge she was now living with a man named Ronny Chisholm down some bad road to an abandoned mine up in Dutch Flat.

From there, the rest of the story came limping in bit by bit, all torn up. A week later I was filling out reports in our kitchen when our part-time secretary — no one called her by her name; she was referred to only as MacGaff's Girl Friday — called on the intercom from the little front office at the lower end of our compound. A woman had come in wanting to talk with me personally and no one else, she said. I told her I was on my way.

When I opened the door to the little anteroom of the office, MacGaff's Girl Friday was bent over her typewriter and Mary Murphy stood staring wistfully at a framed photograph of a meadow full of poppies and lupines hanging on our office wall. She was a stoop-shouldered woman in her late thirties who looked like this crash wasn't the first bad thing that had ever happened to her. Her clothing was asexual — old jeans and a lumpy brown blouse. She wore no makeup. Her face was weathered and plain, and bore an expression of blank-faced sadness you see in women whose main talent in life is getting mixed up with the wrong men.

At the sound of the screen door closing, she turned to look at me. I introduced myself. She said she knew who I was: I was the one who had taken Ricky's gun. I told her I guessed I was starting to be famous. She said she had come only to get her purse. I told her I would release her purse, but first I wanted to ask her some questions about how her car had come to be upside down against a tree in Bunch Canyon.

She looked at me and then away, and said in a low, scared voice that she didn't want to talk about it. Her boyfriend would be very mad if she did. I told her as gently as the circumstances allowed that she might be charged with abandoning her vehicle at the scene of an unreported accident if she didn't. I ushered her through the sliding door between the receptionist's anteroom and the small windowed office where the rangers did paperwork. The other rangers were all out on patrol. I offered her a glass of water. She declined. I motioned to an old chair by one of the desks. She sat down. From a drawer I removed a pad of lined notepaper and placed it on the green blotter in front of me. She kept her eyes lowered and her hands in her lap, fiddling with the keys to the borrowed car outside.

I asked her a few unthreatening questions — where she lived, where she worked (she didn't) — and then began to inquire about circumstances of the crash. Again she said that her boyfriend had instructed her not to talk about it. "He's got a temper and you don't cross him," she said. For a while we went around and around, until finally she said that someone named Ricky had raped her, and that's how her car had come to be over a cliff along Yankee Jims Road. I got up and closed the sliding door between us and the tapping of MacGaff's Girl Friday's typewriter.

"Just tell me the whole thing from the beginning," I said, sitting down again.

She sighed, and began in a low monotone. On Wednesday — the day before we found her car — she had been driving onto Interstate 80 headed east at Auburn when she saw two men hitchhiking on the on ramp, Ricky and his partner, Jerry. She normally didn't pick up hitchhikers, but the one with the red beard — Ricky — gave her a really nice smile, so she stopped. They got in — Ricky in front — and on the way up the highway the three of them made small talk. Ricky told her they were miners, living down on the river. They seemed nice, so she offered to drive them home to Yankee Jims Bridge.

When they got there, Ricky asked her to stay and drink some wine. She did. Eventually they ran out, so she drove them up to Colfax to get another bottle. She was a little tipsy by then, so on the way back she gave Jerry the car keys, and he drove them back down to the bridge. It was getting late, they had some more wine, and after a while she realized she was wasted. So she walked alone down to the riverbank beneath the bridge to sober up. After a while Ricky joined her there. They talked for a while and he began to fondle her breasts. She screamed and pushed him away. She wanted to leave, but Jerry had taken the car keys and passed out somewhere; she didn't know where. So she went back up to the road and sat in her car. In a few minutes Ricky showed up with the keys, got in, and began to drive along the road. A ways up he stopped and said he wanted to make love. She didn't refuse.

"Did he have a weapon? Did he threaten you in any way?"

"No. But he was wearing a sheath knife," she said.

"Were you worried he would use it on you?" I asked her.

"No. He never took it out. But I was still afraid of him."

"Go on."

Well, she said, they had sex there in the middle of the road, with her on her back in the gravel. Suddenly a white van appeared around a turn from the Colfax direction. She jumped up and started yelling.

"What did you yell?" I asked her.

"Like, 'Help, he's going to kill me, he's raping me.'"

"Go on."

The man in the van was Rattlesnake Jim, she said. He bought gold from the miners down on the river. He got out and asked her what was going on. She told him she wanted to leave, and then, emboldened by his presence, she grabbed the car keys from Ricky, got into her Chevrolet, started it, and put it in gear. Ricky tried to stop her by putting his foot in the door as she was trying to close it. They struggled, and she stepped on the gas by mistake and drove off the cliff. She was pretty drunk, she guessed.

The car rolled and came to rest against a tree. She was stunned but unhurt, and Rattlesnake Jim climbed down and helped her back up to the road. He said he was worried about her and offered her a ride home to her boyfriend's house. She accepted, and the two of them left in his van, leaving Ricky to walk back down the road to the river.

"What happened when you got home? Did you call the sheriff?" I asked her.

No, she said, her boyfriend didn't have much use for the police. Later that night he drove her back down to the bridge. When they

got there they found Ricky. Her boyfriend woke him up and made
her tell him if Ricky was the man who had raped her.

"What did you say?"

"I said, 'Yeah, I guess he did,' and they beat him up."

"Were you injured by Ricky during the rape?" I asked her.

"Not really. But my back's pretty sore from lying on the gravel
with him on top of me," she answered. Avoiding my eyes, she
turned her face away.

I picked up the phone on the desk, called the Women's Shelter in
Auburn, and asked them to send a social worker.

"Are you at all sore or injured below the waist?" I asked her, putting down the receiver when I finished the call. "I'd like to have you
examined by a doctor. It won't cost you anything, and we've got to
do it to prosecute Ricky."

No, she said, she hadn't been hurt down there and she didn't
want to see any doctor. I asked her if she had washed her clothes
and underwear. Yes, she had, she said, because they were all dusty.

When the caseworker from the Women's Shelter arrived at the
ranger station, I asked her to chaperone us into the little utility
room where the copier, the fridge, and stationery supplies were
kept. Closing the door, I asked Mary to remove her blouse, leaving
her bra on, and face the other way. She did. I saw some small
bruises and scabby scratches on her back. I took some Polaroids of
them, then told her she could get dressed. Leaving the social worker
with her, I went to get her purse. When I came back, she and the social worker were waiting in the anteroom, talking quietly with
MacGaff's Girl Friday. I gave Mary Murphy her purse and my business card with the Women's Shelter's phone number written on it
in ballpoint pen. I told her I was sorry about what had happened
and to call me if she needed anything. She thanked me, I thanked
the social worker, and the two of them went out the front door together. Outside they got in separate cars and left.

The next day I checked the hospital. There was no record of a female assault victim matching Mary Murphy's description since the
previous Wednesday evening. But on the fifth a man had come in

so badly beaten up that an emergency-room nurse had called the sheriff, and a deputy had been sent over to take a report. I went to the Sheriff's Department and got a copy. The victim was Richard Samuel Marks, his address the North Fork of the American River. In the narrative the deputy stated that although he questioned Marks for some time, the injured man refused to say who had attacked him and why.

I had three days off, and I tried to forget the canyons, the vertiginous bridge, the dark slatey cliffs, the bullet-riddled cars, and the dust. On my first day back, I went to Yankee Jims Bridge looking for Ricky Marks, Jerry Prentice, or any witness to the alleged rape or the beating that followed. I especially hoped to find Rattlesnake Jim, the gold buyer, whom I'd seen several times that summer hanging around the North and Middle Forks in his white van. He wasn't there and neither was Marks, but I did see another miner waist-deep in the water next to a dredge just upstream of the bridge. When he saw me walking down to the riverbank toward him, he emerged from the water and shut the noisy machine off. I asked him if he'd seen Ricky or Rattlesnake Jim. He told me the North Fork was about the last place I'd find either of them, because they were both scared to death of Mary Murphy's boyfriend. I said I could understand why Ricky would be scared, but what did Rattlesnake Jim have to fear from the boyfriend?

"Didn't you hear?" answered the wet-suited man, pulling off his diving gloves and lighting a cigarette. And then he related how, when Rattlesnake had driven that woman home to Dutch Flat, her boyfriend had kidnapped the gold dealer and forced him to lead his bunch of vigilantes back down to where Ricky was sleeping. Rattlesnake Jim had been held at gunpoint and forced to watch while Ricky was beaten within an inch of his life.

By the time the last pieces fell into place the following day I had a queasy feeling every time I looked at the manila folder on my desk with the witness and victim statements and photographs accumulating inside it. After a few years as a ranger, you can tell when it's going to rain from the smell of the air and which way the wind's blowing, and that morning I must have felt the wind blowing in a certain sick direction, because purely on a whim I picked up the phone and called the Colfax Police Department. When the chief answered, I asked him if he knew a miner by the name of Ricky Marks.

"Well, funny you should mention Ricky, because he's sitting right here. He's a real mess, and believe it or not, he's come to seek police protection."

I asked the chief to hold him there until I arrived.

The police station was a nondescript old two-story stucco building on the main street of Colfax, a rough little town nestled in a valley in the pine and red-dirt hills along Interstate 80. Across the street from the station were the rusty tracks of the switching yards, a padlocked Southern Pacific Railroad passenger station, an abandoned freight station, and a four-story wooden hotel with all of its windows broken out. These had been the town's vitality before the lumber mills, mines, fruit-packing outfits, and railroad had consolidated their operations elsewhere. A few doors down from the Colfax police was the Station House Saloon, where the waves of men laid off by these companies had mumbled over their beer and fought with each other for decades, and where Ricky Marks had received the first of his beatings that summer.*

When I came in, he was seated on a gray metal chair with his back against the pale wall next to one of the policemen's desks. He

---

* Colfax is much improved since 1986 as a result of an upswing in Placer County's economy. The Station House (not its real name) has changed hands and is today as pleasant, decent, and safe an establishment as you might find anywhere. On a recent afternoon there, sipping straight whiskey in the dim light from the front windows, I asked a grizzled fellow customer what he remembered about the place circa 1986. He paused, then answered, "The fights, at least two of them, every Saturday night," and turned back to the beer he was nursing. The Colfax police station is now a sheriff's substation, and as this book went to print the Union Pacific Railway was offering passenger service from a newly renovated station and the long-windowless wooden hotel across the tracks was being renovated for occupancy after decades of decline.

was a wiry man of thirty-six in jeans and a T-shirt. His hands were callused and his nails broken and blackened, and he had the ropy arms of someone who moved stones in the river for a living. His face, beyond the regions covered by his thick red beard and hair, was a mess of purple and greenish yellow bruises, black sutures, and crusty dried blood. His eyes were swollen nearly shut.

I introduced myself. He told me he knew who I was. I said I guessed I was becoming famous. He said he was sorry about the shack, but he knew he couldn't stay there anyway because his life was in danger. He and Jerry planned to go up there and disassemble the dredge before winter. I told him to remember to stop by our office and pick up his gun and machete.

I sat down with my notepad, and he told me a story that was pretty much the same as those I had heard from Mary Murphy and the other witnesses, differing in only a few key details, such as the consensual nature of the central act. According to Marks, after the run to the liquor store they'd all gone skinny-dipping in the river, and the woman hadn't objected to kissing and having her breasts fondled, not only by him but by his partner, Jerry, and by Kenny, another miner who had happened along and helped them drink the rest of the wine. Later, when he had driven up the road with her, he said, she had agreed enthusiastically, if drunkenly, to his suggestion of oral sex. They had gotten out of the car and started making out and ripping their clothes off, but they were both so drunk they fell down and finished the act with her lying naked in the gravel. Thinking back on it now, he didn't blame her for yelling like that when Rattlesnake Jim appeared around the turn — the gold buyer knew everybody in this country, and considering how crazy her boyfriend turned out to be, he could see how she might have been worried about the news getting back to him. Maybe they had all had a little too much to drink, but it sure was fun until it turned bad. And then it was bad, real bad.

"What kind of rifles did her boyfriend and the other men who beat you have?" I asked him.

"Like paratrooper rifles — the army kind. Full auto," the miner answered.

"What does he do for a living? Can you show me the road where he lives?" I inquired.

"Do I look that stupid? What do you think he does for a living? Something with drugs, that's for sure," he replied.

When we finished talking, I left the police station and drove back into the canyon to Yankee Jims to look for Rattlesnake Jim. This time I got lucky; his white van was parked by the bridge. I found him down at the river haggling over a small vial of nuggets with the miner who had told me about his kidnapping. Rattlesnake Jim looked nervous when he saw me.

The gold buyer's story pretty much lined up with everyone else's. He really didn't know whether Mary Murphy had been raped or had gotten scared her boyfriend would find out about Ricky Marks. But there was no question her fear of her boyfriend was justified, he said. That boyfriend was one crazy son of a bitch, or at least whatever he and his buddies were doing down at that old mine, they didn't want anyone to know about it. When Rattlesnake had driven the woman home — and it was way the hell down this dirt road where the old-timers had hydraulic-mined the land into tortured hoodoos and the trees were all twisted — the boyfriend had emerged on the front porch of his cabin and started shooting at Rattlesnake's van before he even knew who it was. Rattlesnake jumped out and threw up his hands, begging for mercy and yelling that Mary was in the van. And then, for all his chivalry, he had been kidnapped and forced to lead the boyfriend and his bunch of vigilantes back to Yankee Jims Bridge, where, with one of the boyfriend's buddies holding a rifle to his head, he had been forced to watch as they shoved Mary into the camp and made her identify Ricky and then beat the poor miner in the face and groin with their rifle butts. And when they finished, they told everyone there to keep quiet or they'd get the same thing. And that was about it, said Rattlesnake Jim.

In the end Finch wrote a five-sentence report about finding and impounding the wrecked Chevrolet, but by ranger custom, as the one who'd first seen the tip of the iceberg — that squatter's shack up the North Fork — I inherited the rest of this sprawling mess. I was left staring at my notes and wondering what kind of a raid team — ten men with assault rifles, tear gas, a helicopter, police dogs — it would take to even question the boyfriend and not get shot down in some mercury-laced wasteland of an abandoned hydraulic mine up in Dutch Flat. Four days later I wrote it all up, cut a copy to a sergeant of detectives I knew at the Sheriff's Department, and set up a meeting between him and Rattlesnake Jim. In the months that followed, Mary Murphy declined to press charges, and I doubt that detective sergeant or anyone else ever got around to seeing the boyfriend. I went on to other things and tried to forget the scared looks on the faces of Ricky Marks and Mary Murphy.

When I was growing up in California, schoolchildren were fed a pretty, triumphal account of the Gold Rush. To hear it told, it had been a rollicking good time. Later I learned that for the land, waters, and native peoples of California, and even for most of its participants, the Gold Rush was a disaster.

By the mid-1850s the American River canyons would have been unrecognizable to anyone who had seen them a few years before. Miners had lifted the river out of its bed and put it into miles of wooden flumes so that open-pit mines could be dug in its bed to recover gold. To get wood for the flumes and waterworks, pit shoring, bridges, and temporary towns along the riverbanks, tall forests of pine on the canyon walls and rims had been clear-cut. As a result of these activities, thousands of tons of topsoil were lost to erosion.

With the invention of water cannons to blast gold out of higher ground away from the river — a process known as hydraulic mining — the Gold Rush became a water rush. Mining and water companies diverted hundreds of streams into ditches cut across the canyon walls to the mines. By 1867 all of the miners' aqueducts in Placer and El Dorado Counties, placed end to end, would have stretched from there to Minneapolis. For three decades, hydraulic miners committed mayhem in the Sierra. When it was over, 255 million cubic yards of mine wastes and mud had gone down the American River alone, the equivalent of 25 million full-sized semi dump-truck loads. Over a century after they closed, the hydraulic mines remain — miles of barrens bleeding mercury into the river, like the one in which Mary Murphy's boyfriend was probably manufacturing methamphetamine. After the hydraulic mines shut down, for several decades one-hundred-foot-long bucket-line dredges churned the material hydraulic miners had washed into the beds of the North and Middle Forks for gold they had missed, while above them the canyon walls were overgrazed by cattle, mined for limestone, cut over for second-growth timber, and burned repeatedly by human-caused fire. To make the area safe for cattle and just on general principle, mountain lions, bears, and coyotes were tracked down and exterminated.

In the early twentieth century the water rush continued, now supplying irrigated agriculture, cities, and hydroelectric power stations. Improvements in technology made it possible to build dams that could inundate whole landscapes, and the water and electricity businesses joined forces with the constituency for flood control in poorly sited cities like Sacramento to build them. And so the Gold Rush led to the Auburn Dam and a tradition of valuing what could be extracted from these canyons more than the canyons themselves. Feminist historians have likened it to valuing a woman more for her sexual favors than for her personhood.

However, most of the human victims of the Gold Rush were men — dissatisfied men; men who left their homes and families in other parts of the world and came to the mountains of California wanting something better. As did people like Ricky Marks and Jerry Prentice in the gold rush of the 1970s and 1980s, the original Gold Rush miners suffered from drunkenness, illness, violence, and poverty more often than they prospered. Historians have estimated that only one in twenty made good. Of the rest, the lucky ones went home empty-handed or found other occupations. The less fortunate contracted cholera, malaria, or other diseases and never went home at all.

There isn't much left of all the wishes and hopes miners brought to the American River in 1848 and 1849 but a few platforms on the canyon walls and an abiding wildness in the culture of California. In the 1970s the Bureau of Reclamation hired a team of salvage archaeologists to survey those cabin sites, when it seemed their story would soon be lost beneath the waters of the Auburn Reservoir. Digging the telltale benches along the canyon walls, the archaeologists found a lot of broken bottles of the kind that once held whiskey and patent medicines, for all the physical and spiritual ills attending this rough miner's life on the river. I've read the archaeologists' reports, and what struck me was that not a bit of gold was found in the footings of those camps. The gold all left here for a bank vault in some faraway city. What remained in these canyons was a certain way of looking at land, waters, and women and a hollow yearning afflicting some members of every generation that neither gold, nor sex, nor wine or whiskey can repair.

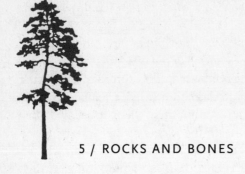

5 / ROCKS AND BONES

WORKING IN THE CANYONS that would be flooded by the Auburn Dam, I couldn't help but become a student of it, in the same way a ranger at a Civil War battlefield can recite the minutiae of Pickett's Charge or the Bloody Road. Soon I knew the story of the dam's political advances and geological defeats, the campaigns to rally its supporters and turn the tide on its critics, and the technical details of what had been done so far in the construction site, which we were required to patrol once a day so the curious wouldn't impale themselves on a piece of rebar or get sucked into the diversion tunnel and drowned. And wherever we went, people asked us questions about the dam.

One autumn morning I stopped in at the Auburn Police Department to get permission to use their gym for one of my defensive tactics training sessions. While I was there, an officer I knew introduced me to their new guy, Rich Morita, who'd just been rotated in off patrol for his first stint at investigations. We made small talk over Styrofoam cups of weak detective coffee, and when Morita learned that I worked in the dam site, he asked me if I had a few more minutes to talk. I said I did. The officer who'd introduced us drifted back to his desk.

I followed Morita across the detectives' office to two battered cardboard file boxes on a counter along one wall. He removed a folded piece of paper from one of them and smoothed it out on the counter. It was a primitive computer-plotted map of the lower North Fork and the western edges of Auburn, with a scattering of red stars on it.

He cleared his throat and began: "I got this from the Bureau. The red stars are old mineshafts."

"Okay. Go on," I said.

"On September 8, 1982, a woman named Karen Dellasandro disappeared from her home here" — he dropped an index finger on one of the squiggles of streets in the Skyridge housing development along the canyon rim — "where she lived with her husband and two kids. He was a sheriff's deputy."

"Right. A sergeant at the jail. I've met him, and I've heard the story — a little of it, anyway," I said.

Morita corrected me: "He's not with the Sheriff's Department anymore. He transferred to the Southern Pacific Railroad Police, but he still lives around here. Anyway, there was a missing persons investigation by Auburn PD, and over the months that followed it began to be looked at as a possible homicide and focused on her husband as a potential suspect. But here's the problem —"

"I know," I interrupted him. "No body."

"Right. No body. She was never found. And no murder weapon or anything else that could bring the case to trial as a homicide."

"So?" I asked him.

"Well," he continued, "during the first investigation they tried about everything they could think of to find her, and around Thanksgiving of 1982 they even consulted a psychic. The psychic told them she'd seen Karen's grave around the arch of the dam's foundation, downhill from those radio towers on the hill by the Auburn Dam Overlook — you know the place, less than half a mile from the Dellasandro home."

"Yeah, I know the place."

"On the fourth of December, the sheriff's search and rescue team assisted our detectives in a ground search of that area with four of their dogs. It was one of several searches they made in various areas of the county around that time, most on hunches and tips and few, if any, on solid leads. Anyway, I guess by that time there wasn't much construction going on, and —"

"But a psychic — I mean, that's really grasping at straws."

"Well," he replied, "it's the closest open land. It's less than five minutes from their house, so he could have left his kids asleep in their beds that first night. The place was deserted once the Bureau people went home at five. And our profiler from the FBI says, based upon what is known about the type of crime, he would have put her body somewhere close where he could keep an eye on her. It's about control —"

Again I interrupted him. "And of course, once the Bureau gets everything worked out with the dam, the grave goes under a couple of hundred feet of water."

"That too." He nodded patiently. "Anyway, the psychic said she saw a grave, a mound of earth, I guess — and in the ground search they didn't find anything recent enough to dig up. But they found three old mineshafts, or what they described as mineshafts, and for some reason — maybe they ran out of time, or they hadn't brought lights and caving equipment — they didn't search them."

"And . . ."

"And there's a note in the report . . ." He rifled through the stacks of manila folders in one of the boxes. "Well, it's here somewhere — saying the PD was going to come back and look in them later. But I've gone through the rest of the reports, and there's no indication we ever did."

"I'm getting your drift."

"Yeah," he said. "That's what I want to ask you. Do you know of any mineshafts on that side of the dam site? They're not on this map the Bureau made for me."

"Well," I responded, "I don't recognize the particular mineshafts you're referring to, but I can tell you that there are old mineshafts all over in these canyons, and a lot of them are not on present-day maps. And the dam site is *full* of things that look like old graves — miles of slit trenches the geologists dug and then filled back in, to study potential faults after the Oroville earthquake. And the rocks under the dam? They're like Swiss cheese . . . there's over nine thousand feet of abandoned underground tunnels, drifts, and raises under the dam site, and over a hundred thousand feet of bore holes they made taking core samples."

Morita was looking at me. "You think she's in there, don't you?" I said.

He shrugged.

At that point a blue-uniformed receptionist stuck her head through the door and called Morita to the front counter to talk to a witness who'd come in about another matter. Morita excused himself. I was left looking at the map of those mineshafts and wondering if I'd been feeling a little down when I'd walked in there, or if it had come on during our discussion. Maybe it was just the time of year.

By September the foothills were worn out from the deprivations of the dry season, and everything waited breathlessly for rain. The dusty trails in our canyons were covered with the riverine tracings of whiptail lizards and racer snakes, and although it was still hot, there was a certain aging of the light. The smoke from the burning stubble on the rice fields down in the valley backed up against the mountains, and the lowering slant of the sun through it brought a nameless melancholy to the mornings of the well-adjusted, and to the desperate, more desperation. What had you done, now that the year was three quarters gone? What had your life come to? What of the New Year's resolutions you'd made? The dry weeds stood in silent ranks on the roadsides, the wind didn't blow the smoke away, and at the Maidu graveyard across Auburn-Folsom Road from the Indian Rancheria at the south end of town, bunches of faded plastic flowers and abalone shells from the coast were arranged on the bare humps of red clay between the dry grass. Many of the graves had only the crudest markers, a name painted by hand or applied in adhesive mailbox letters on a ceramic patio tile, laid upon the mound. Others had no markers at all.

Of course, for most of the time there have been human beings that is all we could expect, an unmarked grave. But we want better than that now. We want order, completion, closure, a granite monument, a public mourning, and in the event of some wrongdoing, an orderly assignment of it to the culpable. So when Morita came back, I offered to help. My involvement in the case would never be anything more than peripheral, but for the rest of that fall I did what I could to help them find Karen.

In the photograph they'd used on the missing-person poster, she was pretty, in her late twenties. Her dark, glossy hair cascaded in curls just past her shoulders, and her blue eyes sparkled with intelligence. Her smile was broad and generous, accentuating her cheekbones and perfect teeth and the little dimple in her chin. Morita told me she'd been a stay-at-home mother, but in the photo she wore a navy blue blazer and a high-necked white blouse with a little lavender bow at her throat. She looked like a job applicant, one most of us would have been pleased to hire.

Skyridge, where she'd lived, was a lakefront subdivision still waiting for its lake. The more expensive homes would have a view of the lake whenever it was finished, but Les and Karen's wasn't one of them. Theirs was what real estate agents call a starter home, an arrangement of shed-roofed lean-tos with an attached garage, sheathed in regrooved plywood stained gray with white trim. It was located at the end of a cul-de-sac where the developer had wrung one more awkward lot out of his acreage. The back of the house was hard up against a steep, wooded hillside, and from the west wall the ground fell away into a swale. In the swale was a barbed-wire fence, and on the far side was a cow pasture, and on the far side of that, the Stations of the Cross on the grounds of the Our Lady of Mercy nunnery. At the south end of the same pasture was the Indian graveyard.

Thirteen years after his wife disappeared, Les Dellasandro went to court to have her declared legally dead so he could sell their home. It had been rented out for years by the time I got involved, and the tenants had used it hard. The untrimmed oaks had grown in around it as if in shame, and the garage door sagged under its own weight and the weight of whatever had happened inside it on September, 8, 1982, when the house was new and Karen and Les's kids were little.

In 1966, after the Bureau of Reclamation received a $425 million appropriation to construct the Auburn Dam and the Folsom South Canal to carry the water away, the agency found out that the dam they'd sold Congress wasn't practical. There simply wasn't enough low-value land in the immediate vicinity from which to quarry the huge quantities of fill needed to build it. A 685-foot earth and rock dam of the kind the Bureau had proposed would have required the strip-mining of five square miles to a depth of 30 feet. But the site was right next to Auburn, the county seat of Placer County, and surrounded by residences, cattle ranches, and orchards. So in 1967, after evaluating more than thirty alternatives, the Bureau announced a new, daring design. It would be the largest double-curvature thin-arch dam ever built, 196 feet thick at its base but only 40 at its over-three-quarters-of-a-mile-long crest, a veritable eggshell of steel-reinforced concrete.

There are two basic kinds of dams: gravity dams and arched dams. The dam the Bureau had completed the decade before at Folsom was of the first type, a clunky, 340-foot-high lump of cast concrete with great dikes of earth and stone on either side, holding back more than 326 billion gallons of water by the sheer mass of its materials. In contrast, a dam like the one the Bureau now proposed for Auburn derives its strength not from sheer mass, but rather from its elegantly engineered shape. An arched dam transfers forces outward into the walls of a canyon in the same way the arched ceilings of Gothic cathedrals transfer their weight outward and down into the cathedrals' walls and flying buttresses.

Because of this, an arched dam's strength ultimately depends on the strength of the rocks into which it's built. Unfortunately, the rocks of the Auburn Dam site were not the best. They were a jumbled mess of amphibolite and chlorite schists, metavolcanics, slick green serpentinite, and talc zones, all extensively faulted and intruded by mafic dikes, quartz, and calcite veins and sloping crazily to the north. Several generations of graduate theses in geology could have been done on them, and indeed, over the following decades, they became some of the most-studied rocks in the world.

But flush with three decades of accolades and swelling budgets since Hoover and Grand Coulee, the Bureau's engineers thought they could repair the flawed rocks. Two years after the thin-arch design was announced, the Bureau finished blasting the first seven thousand feet of tunnels under the massive crescent-shaped engravure they were cutting into the canyon walls. From inside the tunnels and from the surface above, they drilled nearly fourteen miles of core samples, and using these, they made elaborate charts of the extent of the weaknesses. Then they excavated the worst sections of rock and filled the resulting cavities with over two hundred thousand cubic yards of concrete. Not surprisingly, the engineers referred to what they were doing as "dental work," and in dam construction, it wasn't unusual. What was exceptional was the extent of it. At Auburn, it might have been better described as dental reconstructive surgery — like several root canals and a whole suite of crowns.

Regrettably, the site's problems were deeper than the teeth. In fact, the canyon's jaws, indeed its whole body, were a Frankenstein's monster assembled out of bits of older stuff grave-robbed from dead landscapes elsewhere and sutured together at ragged, partially healed scars. And while dam geology concerns itself more with description than narrative, the dam's engineers and geologists had no idea at the time how the rocks they were working on had been formed, and no idea of the nature of the motions even then underway beneath them. From the first surveys of the dam site in the 1920s through the decision to build a thin-arch dam, the American River country had always been thought of — if warped, twisted, uplifted out of the sea, upended, and its rocks remanufactured by heat and pressure — as, in some sense of the word, still a *place*. But in fact the Sierra foothills were a whole collection of places, plucked from a world map that, reconstructed backward at fifty-million-year intervals, looks at first distorted and then entirely unrecognizable.

By 1982, Les Dellasandro was a supervising deputy at the old sheriff's jail next to the county courthouse in Auburn. On his days off he tended a few cattle he kept on the ranch of a fellow deputy over in El Dorado County. In a town where horse trailers vastly outnumbered European cars, that didn't make him unusual. He was tall and blond, with a well-trimmed mustache, and he had a kind of natural gravitas that demanded respect but didn't invite easy friendship. "He's a cold fish," a retired Auburn businessman once told me. "I never liked him much."

Les's wife's nose had been broken a long time ago, either before or just after she met him, and she had a scar on her arm. Karen told people she'd gotten these injuries in an accident, which may well have been true. But in the days after her disappearance witnesses told police that they'd seen Les ridicule her for these flaws in her appearance, calling her names like "Scar Arm." One neighbor told investigators she could always tell when Les was home, because, she said, you could hear the shouting from his house. A fellow deputy told detectives that he and his wife had been seeing the Dellasandros socially, and on a double date at the county fair the previous summer, they'd witnessed Les Dellasandro excoriating his wife, saying she was dressed like a prostitute. The deputy said the incident had made them so uncomfortable they'd stopped seeing the Dellasandros. Another coworker of Les's phoned the police when he heard Karen was missing, telling them he feared for her safety because he believed Les was fully capable of hurting her. But we've all probably heard something similar about some unfortunate couple in our acquaintance, and to be fair, these accounts don't add up to a murder or, for some married people, even a divorce.

From a light aircraft like the one later used to search for Karen's grave, you see a certain visual unity to the west side of the northern Sierra. Beneath you, Placer and El Dorado Counties are a rolling sheet of hills that rise so gradually toward the Sierra crest some fifty miles east — at an overall slope of only two or three degrees — that the landscape appears, if bumpy, almost level. This hilly surface has been cut by deep river canyons into a series of discrete ramps, the even tops of which, lined up by eye as you look north or south, reveal their origin as a whole. But this uniformity is recent, the result of an icing of lava, mudflows, and ash from a period of volcanic eruptions between about thirty and ten million years ago. The swirled marble cake beneath this icing, exposed in the walls of the river canyons, is anything but uniform.

Three hundred million years ago there was no California, and the rocks under my chair legs as I write this were being deposited on an ocean floor off a coast somewhere in what is now Nevada, as layers of silt, sand, and a gentle, steady rain of dead marine life in an ever-changing array of forms — some bizarre, some beautiful, most too small for the eye to see. Underlying these sediments was a thick layer of charcoal-gray basalt, which — as any college student knows today, but the dam's designers didn't — was being extruded as molten rock from a seam somewhere out in midocean and moving toward the landmass that would become North America as it was progressively created.

With all of this movement, a map of the world of about 250 million years ago, as I have said, looks nothing like the world map today. By then the earth's landmasses were joined together into a supercontinent known to today's geologists as Pangaea. Pangaea began to break up, and its coast in the middle of what is now Nevada began to jostle against the sheet of sea floor in what is now the Pacific Ocean, and at various times the ocean floor continued to move toward the continent. The continental rocks were lighter, or so the present theory goes, and tended to ride up over the sea floor as the two collided. The pressures were immense, and in the process of this inexorable collision chunks of ocean floor and whole archipelagos of volcanic islands riding landward on it were scraped off and applied in layers to the growing margin of what would become California.

The poorly healed sutures between these additions to the continent formed zones of weakness, and as the jostling between coast and sea floor continued over the eons, strain built up in the rocks and systems of cracks formed between the layers. Today one such system, the Foothill Fault Zone, runs down the western front of the Sierra in two major strands: the Melones Fault Zone, just east of my part of the American River, and the Bear Mountain Fault Zone, to the west. Crossing the North Fork of the American below its junction with the Middle Fork, the various component cracks of the Bear Mountain — the Salt Creek Lineament, the Maidu Fault, the Spenceville-Deadman, and the southern extension of the Wolf Creek — converge like the waist of an hourglass, as if squeezed between the great iceberg of sparkling granite called the Penryn Pluton to the west of the dam site and the mash of old sea floor to the east of it. That, at the risk of oversimplifying it, is what we know — or think we know — of how this country was formed. And we know that the waist of that hourglass of cracks runs right under the site of the Auburn Dam, and that the cracks may be moving, or probably are. But it's an entirely different story from the one the 1969 geologists had learned in school and on which they based their judgments of the potential for earthquakes in the dam site.

In 1966, the year the Bureau established its dam construction office in Auburn, the California Division of Mines and Geology published what some older geologists in the state now refer to affectionately as "the Old Testament."

Bulletin 190 was a compendium — big and richly illustrated with complex diagrams — of what was known about the geology of Northern California. At the time of its publication, the theory of plate tectonics — which has it that the earth is like a pot of boiling milk and the continents and ocean bottoms so much skin on its surface, made at one location, wrinkled by force, and recycled back into the milk at another — had already been well articulated and supported with a growing body of data. The developing theory was the subject of much discussion in academic circles, but many working geologists still considered it controversial.

Bulletin 190 affords us a look at a widely accepted version of California's geology that was contemporaneous with the thin arch, and entirely excludes what we now see as essential to understanding why the rocks underneath the dam were so worrisome. It's a version that devotes over 49 of the bulletin's 507 pages to the infamous San Andreas Fault and its branches, which caused the 1906 earthquake that devastated San Francisco, but makes absolutely no mention of active faults in the foothills of the Sierra.

Strangely, had that document's eminent authors made a survey of local newspapers, they would have known there had been foothill earthquakes severe enough to scare people out of their houses and ring church bells during their own lifetimes. But geology, like any science, is often directed by commerce and human affairs. After 1906 filled the morgues and cleaned out the insurance companies, coastal geology was about earthquakes (and oil), but in the foothills of the Sierra, it had always been about finding gold. Because foothill faults had moved, but not catastrophically — Auburn's brick business district was still standing sixty years after San Francisco's unreinforced masonry buildings crushed hundreds of their inhabitants — little was known about them, and the Bureau considered them inactive. And a widely held belief in what something — or for that matter who someone — is can be a powerful antidote to the observed truth.

The story assembled later by detectives would show that in the months leading up to Karen's disappearance, the Dellasandros' marriage seemed to be coming apart. One witness remembered Karen expressing fear of her husband, saying he'd recently smashed a chair into bits during an argument. Less than three weeks before Karen vanished, the family dog, a German shepherd named Fuzz, got into the garbage. According to Karen's mother, when Les found out he flew into a rage, flinging the dog around and beating it severely with his fists and feet. The following morning Fuzz was in a coma and Karen took him to a vet's office, where the dog died later that day. After Karen's disappearance the police recovered the veterinarian's medical report. It confirmed that the cause of the dog's death — broken bones and internal bleeding — was consistent with a severe beating. But of course such injuries might be found in an animal that had been hit by a car, too. When detectives went to Les's father's place looking for signs of Karen, they found a smallish mound of fresh earth on his property. Les's father told them that it was Fuzz's grave, and that it contained a blood-soaked quilt the animal had been wrapped in when it was brought there. For whatever reason, the police did not dig it up.

But it was not only Les's behavior that was changing. In her last weeks Karen exhibited new signs of independence and what could be interpreted as attempts to shore up her flagging self-esteem in the face of her husband's withering criticism. She had enrolled part-time at Sierra College and was talking about starting a career. She had confided in a friend that without such an income she could never afford to leave Les and support her children. And she'd begun inquiring about the legal nuts and bolts of a separation and how to get a restraining order against her husband.

Six days before Karen vanished, her mother drove her to a hospital, where Karen underwent cosmetic surgery to remove the scars on her face and arm and had breast implants put in. Her parents told the police they'd paid for that. Karen spent what is believed to have been the last week of her life in postoperative pain, wearing an elastic brace around her chest. The brace and the pain made it impossible to drive, and she'd been relying on her mother and Les to get her around on errands. Investigators came to believe it was highly unlikely she would have picked this time to leave Les but instead would have waited for a week or two, until she felt up to traveling. She'd made at least one previous attempt to leave, but that time she'd taken the children with her, and people who knew Karen well said she'd never have left them with Les.

In the course of the investigation, it became clear that Karen's mother disliked Les intensely and made no secret of it. If even a fraction of what she recounted about her son-in-law was true, she may have had ample cause. But for the sake of fairness, we must remember her prejudice when we assess her testimony. With that said, Karen's mother told police that during a car ride after the surgery — and she claimed Karen had told her this — Les had driven over a bump and Karen had complained about the pain and asked him to take it easy. At this, the mother said, he caused the car to jerk repeatedly, alternately applying the brakes and the accelerator. And, according to the mother, in the days after the surgery he added a new invective to the stream of verbal abuse directed at his wife: "Old Falsies."

After World War II, dam know-how was a key component of United States foreign aid, and Bureau of Reclamation engineers regularly functioned as overseas advisers to developing nations. By the 1960s big dams were going up all over the world. One of them, the Kariba, in Africa, was completed in 1961. At that time geologists considered the region along the border between Rhodesia and Zambia seismically stable. But as Kariba filled, the area was shaken by a series of earthquakes, which increased in severity to a particularly strong set of shocks in 1963, just as the dam reached capacity.

Later, by the mid-sixties, India finished a large dam at Koyna, in a region one report called "one of the least seismically active places in the world." During its filling, mild earthquakes began to occur, culminating in a magnitude 6.4 shock on December 10, 1967. The dam cracked but survived; however, nearby communities were not so lucky. The collapse of unreinforced masonry buildings in those villages killed at least 177 men, women, and children.

In Greece, the rapid filling of the Kremasta Dam after its 1965 completion was contemporaneous with an earthquake that caused slumps and landslides, damaged over 1,600 buildings, killed one person, and left sixty injured. Similar patterns were noted at a French dam in 1963 and a Swiss one in 1965. And by January 1972, the National Academy of Sciences and the National Academy of Engineering released a joint study stating that under certain conditions, large reservoirs had a potential to cause earthquakes as a result of the immense pressures they exerted on what were now understood as discrete portions of the earth's movable crust and, some geologists believed, by injecting the faults with water under pressure.

As investigators reconstructed it, the final rupture in the Dellasandros' marriage occurred around the question of where their children would go to school.

Les was Catholic, and the children were currently attending St. Joseph's School in North Auburn. But Karen wanted them moved to Forest Lake Christian School, in Lake of the Pines. The day before her disappearance she went to Forest Lake Christian to have a look around. An acquaintance later told police she had run into Karen in the schoolyard and had talked with her briefly. She remembered Karen saying that the pain from her recent surgery was considerably worse than she had imagined it would be and that she was very worried about her marriage. The next morning at around eight, Karen phoned Forest Lake Christian and made an appointment to enroll her children later that day. She never showed up.

By that morning, the dam site in the canyon below Les and Karen's house was a great hole in the earth like a strip mine, but it was now strangely quiet there. Gone was the noise of construction equipment; there was only the sound of the river at the bottom. Weeds came up on construction roads along the canyon walls. On the manmade cliffs of the dam's keyways, wooden catwalks and ladders on which workers had swarmed like so many ants were going gray and splintery in the sun. Underneath, the tunnels were abandoned, their interior walls still strung with black rubber electrical cord and, every few feet, a light bulb in a wire shield. But the power was shut off, many of the bulbs were now broken, and in the dark the tunnels echoed with drips of water. The water trickled along their floors, emerging at the tunnel mouths high on the keyways, where willows, their seeds carried there on the wind or by birds, were taking root and would soon grow to hide the tunnel portals entirely.

In the Bureau's palatial offices on the canyon rim, janitors still kept the floors perfectly polished, and each morning the engineers showed up for work as usual. Within another few years one dropped dead at his post from a heart attack, still waiting to finish his dam. NO TRESPASSING signs were posted on all the locked gates into the dam site, and the gates were checked every day or two by the rangers. But they would never have known if someone had let himself in with a key. And it seemed that everyone in some semiofficial capacity — game wardens, utility linemen, volunteer firemen, even sheriff's deputies — had one. But that's just one of many possibilities.

The quiet in the dam site had come about in the context of two major shifts in public attitudes during the 1960s and 1970s. The first was the advent of widespread public concern for the well-being of the natural world, the second a growing disenchantment with government.

Nineteen sixty-two, the year the Bureau of Reclamation distributed its prospectus on the Auburn Dam to Congress, also saw the publication of Rachel Carson's *Silent Spring*, a book whose wide readership is often cited as the beginning of a broad-based environmental movement in the United States. That movement can be said to have come of age with the 1969 passage of the National Environmental Policy Act. Under the new law, the Bureau was required to publish an accounting of the environmental effects of its dams. Within a month of the 1972 release of the final environmental impact statement for the Auburn Dam, the Natural Resources Defense Council and the Environmental Defense Fund filed suit against the Bureau. A federal judge subsequently found the Bureau's statement inadequate and ordered the agency to amend it. By the time the Bureau presented the amendment to the public in August 1973, it was not the same public at all that had approved of the Bureau and its dams back when the project began.

By April 1973 America's leaders had succeeded in sacrificing the lives of fifty-eight thousand young men to a war they'd just lost in

Vietnam. In May the Senate established a committee to investigate President Richard Nixon's rigging of his own reelection, a string of executive office misdeeds known collectively as Watergate. As Watergate unraveled, the Bureau was trying to make its case to the public that drowning the American River canyons was necessary and right, and after a decade of war and domestic turmoil, the federal government's credibility was in sorry shape.

By the end of that decade Americans' prevailing mood toward government was reflected in a reassessment of government's right to tax and spend their money. Big federal dams had originated under the free-spending New Deal in the 1930s, and although the Bureau always made a show of economically justifying its projects, the dams often failed to live up to those justifications. In 1980, Ronald Reagan was sent to the White House promising to cut government and government spending. By the early eighties the price of finishing the Auburn Dam had swelled to an estimated $2.1 billion — too much, thought the Reagan administration and many members of Congress, for something so controversial. Still, the dam might have been a fait accompli by then if not for the Oroville earthquake of 1975.

Oroville was a Gold Rush–era town about forty-five miles north of Auburn, near a strand of the Foothill Fault Zone called the Cleveland Hill Fault. In 1967, the state of California finished building a dam on the Feather River there. It was the tallest dam in the nation when it was completed, a monster pile of rock fill 770 feet high and well over a mile wide, designed to send water through a system of canals and pumps all the way to greater Los Angeles. The Oroville Reservoir took years to fill, and when it did, its 3.5 million acre-feet of water pressed down on the earth with a weight of about 4.72 billion tons.

Unlike most cases of what was now being called "reservoir-induced seismicity," the Oroville event came not during the reservoir's filling but during a rapid drawdown in the dry summer of 1975. On the afternoon of Friday, August 1, a series of violent shocks rocked the area, radiating from an epicenter on the Cleveland Hill Fault just southwest of the dam. The largest reached a magnitude of 5.7. In Oroville, sidewalks and streets buckled and cans and bottles rumbled off store shelves, forming heaps in the aisles. People ran out of homes and businesses into the middle of streets, where some stood thunderstruck, frozen in fear. Others fell to their knees and began to pray, sure that the day of reckoning had come. Schools and county offices were damaged and had to be closed. At a local fire station, firemen watched spellbound as a tank truck full of water jumped up and down on the concrete floor of one of the truck bays. Then the dispatch speaker began to crackle with multiple calls. One man had suffered a heart attack as the ground shook underneath him. Another man had been driving down the main street of town when he looked over his shoulder to watch the buildings wobbling around crazily and collided with a parked car. Falling power lines ignited multiple grass fires. Chimneys fell, or teetered and had to be pulled down.

By September geologists from the U.S. government and the state of California swarmed into the foothills, and as the similarity of Oroville's and Auburn's locations on the Foothill Faults became clear, work on the Auburn Dam was suspended until further geological studies could be completed. Seven months later, the Association of Engineering Geologists' Seismic Hazard Committee released a report saying that even in the case of a moderate earthquake, the Auburn Dam might fail and unleash a wall of water on California's capital and two air force bases downstream. Then no less an authority than the U.S. Geological Survey said the dam was unsafe. The Bureau hired legions of consultants, among them the San Francisco geologists Woodward-Clyde Consultants, to conduct an exhaustive review of the earthquake potential at the dam site. Woodward-Clyde's work was so comprehensive that a quarter century later, quotes from it are the boilerplate of earthquake risk assessments for new structures in the Sierra foothills. When it came out in 1978, the bad news essentially finished off the thin arch — but not the dam. By 1979 the Bureau was back at the drawing board.

The morning after Karen's disappearance, her husband called the dispatcher at the Auburn PD and left a message for one of the policemen to call him. Sergeant Sam Russell, the son of a Placerville logger, returned the call, reaching Les at his office in the jail.

Reflecting on that conversation, Russell later remarked to the captain in charge of the investigation that Dellasandro's voice seemed strangely calm and emotionless as he reported his wife's unexplained absence. During the call Russell collected a few basic facts for a missing persons report, but toward the end of their conversation, he said, Dellasandro asked him not to file the report yet, saying he wanted to look for his wife himself for a couple of days first. In a gesture of professional courtesy, Russell agreed. So it was another two days — three since Karen's failure to show up at Forest Lake Christian — before an investigation started. When it did, the police initially focused on the possibility that Karen had deserted her husband and children and didn't want to be found. They checked local motels, women's shelters, buses, taxis, and the Sacramento airport. But as Karen's friends, neighbors, and parents were interviewed, the investigators gradually changed their opinion of what sort of case they were working on.

Perhaps at that point the circumstantial evidence was too weak to get a search warrant, and maybe the collegial relationship between the Auburn Police and the sheriff's office resulted in a retraction of the usual vigor with which potential crimes are investigated when the suspects are dressed like bad guys instead of police sergeants. Whatever the case, it took several days for the police to get to the Dellasandro residence, and when they did, Les agreed to let them in. They found the house spotless and Karen's jewelry, including her wedding and engagement rings, neatly laid out next to the bathroom sink. Her wallet, cash, and credit cards were all at the house. Investigators checked with the credit card companies for recent activity. None of the cards had been used. For whatever reason, photographs were not made of the home's interior, or if they were, the photos have been lost. Nor did investigators pull up the carpet or pull off the baseboards to look for traces of blood that might have eluded a cleanup. None of the family's three cars and trucks were processed for trace evidence, nor were their carpets pulled up. With 20-20 hindsight, it's clear that the best chance the Auburn police had to gather forensic evidence may have been at the house, where Karen was last known to be alive — her mother had dropped her off there after an errand the night before, and that morning Karen made the call from home to Forest Lake Christian. But it was a consent search, and with no warrant and nothing but circumstantial evidence that a crime had even been committed, such destructive thoroughness might have been justified only by an incriminating statement. However, during repeated questioning, Les Dellasandro stuck to his story.

Here is what Les told the police: In their home the morning of September 8, after the kids had gone off to St. Joseph's, Karen's telephone call to Forest Lake Christian precipitated an argument. In the course of it there was talk about a marital separation. He left the house to go shopping at Kmart and when he returned, Karen was gone. When she didn't come home later that night, he assumed she'd gone to her mother's, so he didn't report her absence until the next day. He couldn't provide sales receipts or any of the things he purchased to substantiate his Kmart alibi, and a check of the store's security cameras proved inconclusive.

Auburn didn't see many homicides, and looking back on it, some of the investigators admit that much has been learned about interrogation techniques since 1982. In retrospect, some of them think that at certain points, had they known some of the psychological tactics they know now, they might have extracted a confession. But that's just their opinion.

It is unlikely that anyone in law enforcement was protecting Dellasandro. Auburn's police chief and officers were known as ethical and hardworking, and they pursued the case diligently for months. They asked for and received the help of sheriff's detectives, who had more experience in homicide, as well as the FBI, the California Department of Justice, and the National Guard, which provided aircraft with heat-sensing instruments capable of detecting a decomposing body from the air. They examined Karen's mother's phone bills to rule out the possibility that Karen had fled with her parents' complicity. They flew an expert in polygraph examination up from Southern California, but Les refused to take the test. They searched the Dellasandros' vacation cabin and the ranch where Les ran his cattle, and for months, the chief told me ruefully, they had a backhoe out digging every time someone in the county saw buzzards circling.

The reopening of the case in which I had a small part was one of several in recent years, and not a whole lot came of it. There was no more reason — other than proximity and opportunity — that Karen's body would have been in the dam site than anywhere else in the three to four thousand square miles of mountains Les knew better than most, having once worked for my own State Parks Department and the Forest Service, even before he patrolled much of that as a deputy.

Karen has not been seen or heard from now for over two decades, and the case sits like a permanent wound on the idea of justice in the minds of everyone who ever worked on it. The evidence custodian at the Auburn Police Department keeps a framed photograph of Karen on a shelf next to her desk and beside it, a bouquet of fresh flowers. Les was never charged and still works as a policeman. He is close to the age now when he can retire on a pension. Karen's mother was so demonstrably bitter toward him that Les eventually got a restraining order prohibiting her and Karen's father from having any contact with their grandchildren. Karen's parents have grown old and the police have taken samples of their DNA, because they probably won't live to see Karen found. For my part, I am haunted by what little I had to do with the case, because cannot seem to separate it from the feeling I now get whenever I'm around the yawning hole of the dam site.

Just over five years before the National Guard's search aircraft flew over Placer County looking for Karen's grave, similar flights were made by Woodward-Clyde Consultants. That time, they were looking for alignments of topography called lineaments — lines of springs, straight valleys, ridges that suddenly zigzagged, or any other sign of faulting. From the dam site they traced these lineaments south down the front of the Sierra as far as Stanislaus Table Mountain, a volcanic flattop split in two and left like a broken child's toy. To the north, they traced them past Les's jail and up the De Witt Lineament past the court where he would get his restraining order and the judgment allowing him to dispose of his wife's property. From there they followed them north through Spenceville, where the shattered hills had been used for artillery practice during World War II, to the Cleveland Hill Fault, which shook Oroville.

After locating these faults from the air, Woodward-Clyde's geologists dug trenches across some of them on the ground. The point of this was to discern — by looking at soil layers displaced upward, downward, or along the faults — how recent and severe earth movements had been in the last hundred thousand years.

Of the forty such trenches Woodward-Clyde geologists made from Lake Oroville to well south of the Auburn Dam, two run under the beginning and aftermath of the eighth of September. One, which Woodward-Clyde called "St. Joseph's Exploration," was dug behind the Catholic school over which, in Les's account, his final argument with Karen had occurred. The other excavation, which they called "Radio Towers," was made below the antennas where the psychic later directed the Auburn Police to look for Karen's grave. This begs the question: Did the psychic see a grave, or was the malevolent mound of earth she was visualizing a fault trench?

In China one branch of earthquake prediction research has focused on the behavior of animals and people, which by anecdotal accounts seems to change in response to strains accumulating in the rocks along faults. Even now, the largest well-drilling company in Placer County still employed two full-time water witches, men who walk through the knee-high meadows of August and September holding copper rods in their hands until the rods wiggle in a certain way, and then drill there. Admittedly, cracks in the ground and water witching are of little relevance to finding Karen Dellasandro, and these details would interest only poets, park rangers, and anyone else who thinks about the mysterious connections between the land and the people on it. But to such as us, the blood that ran in Karen Dellasandro's veins and the cold water in the ground and the water in the American River are all of the same stuff. Everyone in these hills comes from the ground, and we will all return to it, and maybe for the brief period when we are walking around on it, whether with copper rods, murder weapons, or shovels in our hands, the connection between us and it is stretched but never really broken. And so we know things about the rocks we cannot say except with copper rods, and we yearn to have returned to us people whose presence we feel in the rocks and soil beneath us, and in a story told in whispers around town, like the wind that always comes in late September, just before the first rains bring life back aboveground.

ROBERTA WAS A SHORT, round woman in her mid-fifties who worked the swing shift at our dispatch office in Folsom. She liked to remind me that she had been dispatching for the Highway Patrol the year I was born. When I phoned to ask her to fax a printout on a stolen car or some miner with a price on his head, she would answer, "Radio," as if the word had all of the cutting-edge potency of "biotechnology." Her short hair was copper red and always neatly coiffed. She wore sleeveless floral tent dresses and wouldn't have thought of coming to work without her pale makeup on. Her musical tastes ran toward Ella Fitzgerald and Diane Schuur, and she liked to go to jazz clubs in San Francisco whenever she could.

Roberta and I had our go-rounds, but on a moonlit midnight on some dusty road deep in the foothills with the whiteleaf manzanita crowding around you like an army of ghosts, Roberta's voice on the two-way radio was a dry-martini, worldly-wise comfort. One warm night I was on my way home from Sacramento and stopped at the district office below the dark bulk of Folsom Dam, where the aluminum-sided dispatch office sat under the radio mast in the back parking lot. I was walking from my Jeep to the main building when I saw her silhouetted in the doorway at dispatch with a glowing cigarette in her hand. She was humming some smoky ballad over the trebly chatter of the radios. It sounded good, and I remember thinking she could have been a jazz singer. But somewhere she must have taken a wrong turn, and she ended up in front of the wrong microphone.

All of us here felt that way at times. We rangers could have been guarding some jewel-like national park celebrated in expensive coffee-table books. Instead, we would spend years in these purgatory canyons up into the hills from Roberta, where our fellow creatures — black bears, pileated woodpeckers, foothill yellow-legged frogs, Sacramento squawfish, Sierra garter snakes, and pipevine swallow-tailed butterflies — were consigned neither to the heaven of a national park's perennial protection nor immediately to the cold hell of inundation.

Most days I just tried to live in the present. When the morning sun warmed the bark of the ponderosa pines in front of our ranger station and the trees filled the air with the scent of vanilla, it almost worked. At the bottom of the canyon the river still ran deep, cold, and clear as an angel's harp, and in late spring the does emerged from the surrounding woods with their new fawns to nibble at the lawn around our flagpole. They knew us as their protectors, for as any ranger will tell you, animals have an uncanny ability to discern the boundary of a park from the dangerous lands around it. And so you contented yourself with saving things, if not forever, for now. And you tried not to think about where you might have been had things been different, or where Roberta would send you next.

One day Roberta radioed Doug Bell at the ranger station, where he was catching up on reports. There was a ball game on the little clock radio over on MacGaff's desk, and Bell cocked his head slightly toward the announcer's lowered voice as Roberta's first call came in. The bases were loaded with two strikes, the announcer was killing time with a couple of quick statistics, and the pitcher was probably looking at the ground while fingering the ball and shuffling his cleats, then glancing up past the bill of his cap with a measuring gaze at the batter. Bell loved baseball. Each spring he'd take time off, put his wife and kids in his cabover camper, and drive down to Arizona to watch the Cactus League exhibition games.

Roberta's second call ran over the crack of the bat and the rising hiss of the crowd. Bell exhaled, kicked his squeaky swivel chair back from the perpetual disorder of his desk, and walked into the anteroom by the front door, where he leaned over the microphone on the receptionist's desk. Roberta gave him his marching orders — no one else piped up and volunteered — so Bell shambled through the screen door, across the shady porch, and down the concrete path to his pickup. He got in and slammed the door violently, started the truck, jammed it into reverse, chirped the tires, rattled down the driveway onto Highway 49, and turned left toward the river, where he was supposed to meet the victim of a Penal Code 245, an assault with a deadly weapon.

The road Bell followed wound down the wall of the North Fork into the bowl where the North and Middle Forks come together. That junction of two canyons and the need to cross the river where it was easiest had directed wanderers through the Confluence for a long time. At the end of the nineteenth century, a couple of local boys found a hole in the ground near there, which later afforded paleontologists a random sample of foot travelers since the Ice Age — the skeletons of extinct ground sloths, giant bison, saber-toothed cats, and prehistoric people who had fallen into the system of limestone caverns it let into. Before the Gold Rush, Indian trails came to the Confluence from various directions, and a Nisenan village called Chulku had stood there. When the miners finished with it, nothing remained of Chulku but twenty-four mortar holes chipped into a bedrock outcrop by the river, where the inhabitants had ground acorns into their staple meal.

After the Gold Rush, wagon roads had replaced footpaths. As Bell reached the river, the yellow gores of the old roads' switchbacks were visible against the gray-green forests on the canyon walls above him. In front of him, the abutments of generations of bridges littered the banks where floods had left them. The concrete deck of one bridge, from the flood of 1964, lay in the water. Three others remained standing — two for roads, one for a railroad, now abandoned — and high above, a fourth, the new Foresthill, arched across the sky.

From the Confluence in the early 1970s you could watch the two ends of the New Foresthill Bridge grow out from either side of the gorge. As cranes lowered steel to the ironworkers on the two tips, they arched out over the canyon, then down to alight briefly on two towering concrete pylons, and then out again for over 800 feet to join over the river. When the structure was completed in 1973, its twin lanes of pavement ran 2,428 feet along the top of its three arched trusses, so far above the river in the middle that the Washington Monument could have stood upright beneath the bridge with enough room left over to fly a large helicopter between the two.

The featured speaker at the dedication ceremonies that September was Congressman Harold "Bizz" Johnson, an enthusiastic supporter of public works in general and these in particular. He was said to have marked up the House bill authorizing the Auburn Dam — and thus this bridge — over drinks with a local banker and the publisher of the *Auburn Journal* in an Auburn bar called, of all things, the Sierra Club. When he finished speaking, the ribbon was cut as a donkey and an elephant were led out to face each other on the bridge deck — a symbol of the cooperation between Democrats and Republicans that built the new span. Then the crowd bowed their heads for a blessing by Father Brennan of Saint Joseph's, the Placer High School band played a march, and the dignitaries led the crowd in a tour of the bridge.

But without the reservoir beneath it, the bridge immediately became something it wasn't intended to be. Within minutes a man appeared from the crowd and leaped over the railing. He was wearing a parachute, which floated him safely to the canyon floor. Within a year and a half the feat was repeated with a hang glider. Then, on October 8, 1975, a despondent seventy-seven-year-old man from Citrus Heights became the first to jump without either device. Just over a decade later John Carta rode his motorcycle off the span, and in December 2001 a stuntman drove a car off it for the Hollywood feature film *XXX* and, like Carta, parachuted to safety. This time the rangers had issued a permit for the stunt, but it was nevertheless unpopular with them. "No one told us Corvettes are made of plastic and shatter into hundreds of tiny pieces on impact," one of them later remarked. "We're still picking up little red bits of it down there." Meanwhile, the suicides continued.

One man jumped too close to one side of the bridge and survived for a while before he was found dead. With his broken arms and legs, he had made what Ranger O'Leary later told me was an impression like a snow angel in the dust beneath the bridge. Another man wasn't taking any chances, so he sat backward on the bridge rail with a shotgun in his mouth. And in the summer of 2002 a chase involving several police cars ended at the bridge, where the suspect jumped out of his car and into the canyon; he had been paroled from prison and didn't want to go back. It was the second suicide there within a week.

In July 2003, after one man splattered himself on the canyon bottom and another was talked out of it by police, an outraged editorial in the *Auburn Journal* scolded the county — to which the Bureau had turned over the bridge when it was finished — for not doing something: a net, suicide hotline phones on the pedestrian walkway, something. "Unacceptable," the paper said, "that's the only way to describe the fact that repeated calls . . . for action at the Foresthill Bridge continue to go unheard." But the situation was hardly the county's doing. For thirty years the bridge had waited for the water to rise beneath it and stop the carnage, and meanwhile its notoriety had spread to the adrenaline-addled and serotonin-deficient all over the United States. When they got there, someone usually saw them standing at the railing and called 911. The 911 operator called Roberta, and Roberta radioed the rangers.

For all those years there had been another, even more common problem with the bridge sans reservoir: People enjoyed throwing things off it, just to watch them fall. Driving along the river now, Bell saw his man waving from the road shoulder ahead and stopped. The thirty-eight-year-old male victim identified himself as John Geary and introduced his female companion, forty-year-old Lynn Parker. They'd been a little down on their luck, Geary told Bell, and they'd been living in a Roseville motel room and driving up to the river during the day to pan gold. It was a common story. For a century and a half people had been coming to the river in hopes their luck would change. It seldom did. We rangers called them "pilgrims."

Geary and Parker had left their car at the Confluence that morning, they told Bell, and hiked up the river underneath the big bridge. Overhead, cars and log trucks made ominous booms and clanks at the bridge's expansion joints, which echoed through the canyon. Maybe it was these noises or the sheer majesty of the structure, but when he got right underneath it, Geary stood looking up at the bridge. It was then that he noticed three people, mere colored specks, standing at the railing far above. Then he saw them throw something into the canyon. As soon as they let go of it, the object seemed to bloom bigger. He and Parker watched it drift toward them, and as it grew larger they could see it was a little yellow parachute with something hanging from it. Something that seemed to move on its own, as if alive.

A few seconds later Geary could make it out clearly. It was a chicken. A chicken on a little yellow parachute. It floated past them and landed below them in the brush. They stumbled toward it, down the steep talus. When they got to it, the chicken was squawking miserably, hopelessly tangled in its parachute shrouds. So they freed the bird and it fluttered off, clucking indignantly, into the manzanita.

Now Geary heard distant shouts from above. He couldn't make out the words, but the tone sounded angry. Looking back on it, he realized the chicken's owners were mad at him for letting it go. But all he was doing was trying to help. You would have done the same thing, he told Bell. Then he and Lynn Parker heard a series of evil, whirring zips followed by loud cracks, and looking up, he realized that the people on the bridge were dropping rocks, pretty big ones, on them. He could make out the rocks as they were released, but then he'd lose track of them until they appeared again just a couple hundred feet overhead, at which time he and his girlfriend had only a second or two to evade them. They began dodging the rocks, running around like crazy people. The rocks literally exploded on impact. Then Geary saw one of the people above heading for one end of the bridge — to get more rocks, he thought — and he and Parker started running and didn't stop until they were back at the Confluence.

"What did you do then?" asked Bell.

Well, said Geary, they got into their car and drove up the bridge to find the assholes and give them a piece of their minds. But the chicken's owners were gone when they got there, so they went to town to call the sheriff, and then back down to the river to the spot where the dispatcher told them to meet the ranger.

Bell finished his notes. He said he was sorry about what had happened, and that we'd keep our eyes out for anyone walking onto the bridge with either rocks or a chicken. Then he excused himself and left. On the single-page report he wrote back at the ranger station, the last sentence was "No further action." What could you do? "One of them had a bright-colored jacket" isn't much to solve a crime on, and like almost every other problem he had along the river, this kind of thing would keep happening until the reservoir filled.

A few months later it was my turn to go to the bridge. It was early morning, and the sun rising over the mountains on the eastern skyline was painting the west wall of the North Fork canyon in gold and long shadows. I'd picked up some coffee from the pastry shop in town and I was carrying the paper cup from the Jeep into the ranger station when the dispatcher called my number.

"One seven nine, Northern."

"Northern, one seven nine," I answered into the microphone clipped to my epaulet.

"We have a report of a jumper at the Foresthill Bridge, can you respond?"

I pushed the mike button again. "I'm en route."

The quickest way to the bridge took me up out of the canyon and along the rim through Auburn. Siren yelping, I left the gas stations and fast-food joints at the north end of town and rolled down the long straightaway to the bridge. When I got there, I could see a red pickup parked in the opposite lane toward the far end of the span. I crossed the bridge going about eighty, made a quick U-turn, and pulled up behind it. The truck was blocking traffic, with the driver's door open. To my right, three people stood peering over the railing. One of them, a blond, muscular man in his thirties — a construction worker, or maybe a mill hand from Foresthill — left the railing to meet me.

"I reported it," he said as I got out of my Jeep. "We were coming across the bridge when we saw the truck. We saw him get out and step over the concrete wall, then up on the railing. Then he jumped off. He didn't even stop to think."

I peered into the abandoned truck. It had matted gray fake-fur seat covers. The keys were still in the ignition and the radio was on — a country station, playing a slow ballad with steel guitars. An open can of Budweiser, still dewy, sat in the drink holder by the gearshift.

The blond guy looked in over my shoulder. "When I got here just seconds after he jumped, the radio was playing like it is now. But it was a sad song, even sadder than this one — I think that's what made him do it right then."

"Could be," I said. "Some of those country songs are pretty sad." I got in to move the truck out of the way of the traffic now stacking up behind my Jeep. He had been shorter than I was. I moved the seat back and turned the key in the ignition. The interior smelled like beer, cigarettes, and the stale sweat of someone who no longer existed.

As the Foresthill Bridge neared completion in the summer of 1973, all over the country — in airports, at displays of televisions in department stores, in bars, any place where a TV was on — people sat riveted to the hearings of the Senate Select Committee on Watergate as the idea of loyalty was turned on its head. Back in 1959, when the bill to authorize Auburn Dam was introduced in Congress, loyalty to flag and president had been the gold standard of patriotism. But those who clung to it now were cast as villains, like Nixon's chief of staff, H. R. "Bob" Haldeman, who tried to shield himself and his boss from culpability by responding more than a hundred times during questioning that he couldn't remember. The hearings revealed that in addition to burgling the Democratic Party offices, Haldeman and his associates had ordered a break-in at the office of Daniel Ellsberg's psychiatrist. Ellsberg was a government employee with a high security clearance who had turned over secret documents detailing the government's dirty dealings to the *New York Times*. Nixon's men had pried open his psychiatrist's file cabinets in hopes of finding something to discredit him.

This, then, was how much attitudes changed during the construction of the Auburn Dam. Had Ellsberg done what he had in the 1950s, he might well have been put to death for treason. But in the 1970s he became a hero to many Americans, and it was his persecutors in the government who went to prison. So it was in this climate of sympathy toward acts of conscience, as the House Judiciary Committee moved toward impeachment of the president in early 1974, that an Auburn Dam whistleblower emerged from the ranks of the Bureau of Reclamation.

George C. Rouse was no wild-haired environmentalist. A small, sharp-faced man with glasses and a pocketful of mechanical pencils, he was passionately dedicated to his work as an engineer for the Bureau. As a young man he'd worked on the Hoover Dam and when the Auburn Dam came down the pipe, he was attached to the

Bureau's design shops at the Denver Service Center. He soon found himself at odds with his superiors over the dam on the American. Politically, of course, he believed in Auburn and dams in general. It was the thin-arch design he disagreed with. He thought it was dangerous. He would be remembered as stubborn, even combative, and was not the sort of man to be pressured into changing his mind. Finally, in June 1972, Rouse retired in protest over Auburn and went home to his little white frame house on Pierce Street in the Denver suburb of Wheat Ridge.

That winter, as usual, Rouse's car stayed out in the snow. Never one to stop working when he clocked out, Rouse had converted his garage into a home office. For years he'd been excusing himself each evening after dinner to go out there and work on the Bureau's calculations late into the night, and his retirement didn't change that. For the next year and a half, Rouse kept going over the figures for Auburn. And he ordered up some letterhead: George C. Rouse, Structural Engineer.

In February 1974, Rouse typed a letter on that stationery to Harold G. Arthur, the Bureau's director of design and construction at the Denver Service Center. In reasoned, unemotional prose backed with fifteen peer-reviewed references, Rouse refuted the Bureau's whole rationale for the seismic safety of Auburn Dam. Using the agency's own numbers, he derived that the dam could be expected to crack within the first seven seconds of a bad earthquake. He also differed with the Bureau on what he considered its wild optimism on the strength of its concrete, and further — and this turned out to be the most lethal to the agency's credibility — on whether an earthquake could happen closer than fifty miles from the dam site. The Bureau had dismissed that possibility, but Rouse thought it ought to be planned for, because dams last a lot longer than geological opinions. In a later letter he reminded Arthur that people had lived around Koyna, India, for a long time, so there existed a written and oral tradition of what had happened there for the last four hundred years. Nowhere in that tradition was there a legend of anything like the earthquake that had cracked the Koyna Dam and killed 177 residents of nearby villages. Could the Bureau really be so sure that the faults around Auburn would not behave similarly when the dam was filled?

It would bother Rouse for years that he helped the environmentalists, because he disagreed with the Bureau not about dams in general — he loved them, they were his life — but about what he saw as sloppy engineering, with the lives of thousands downstream at stake. Nevertheless, his timing was fortuitous for environmental groups. California had only recently had a wakeup call: two years before, the San Fernando Valley north of Los Angeles was shaken by an earthquake that killed fifty-eight people, damaged thirty thousand buildings, and caused the crest of the Van Norman Dam above the densely urbanized valley to collapse. The dam didn't fail entirely, but its perilous condition caused the evacuation of eighty thousand residents, and thereafter seismic risk had become the darling of environmentalists fighting large engineered structures. They used it against the Diablo Canyon Nuclear Plant on California's central coast, although unsuccessfully, and once the Rouse letters found their way into their hands, they used it against Auburn. What would make Rouse's assertions most damaging was that he was so soon to be proven right. A little over a year after his first letter, the Oroville earthquake shook the supposedly inactive Foothill Faults beneath Oroville Dam.

For the Bureau of Reclamation, the following year, 1976, was a low point. Even as it defended the design of Auburn Dam after the Oroville earthquake, the agency was filling another new dam in Idaho. The 305-foot earth-fill Teton Dam had been completed on the river of the same name the previous fall, and like Auburn's, the rocks underneath it were suspect. During construction the Bureau poured over five hundred thousand yards of concrete into the underlying rocks to keep water from seeping through cracks and voids in them and liquefying the earth of the dam. Teton filled quickly with runoff the following spring, and on June 5, 1976, at five minutes to noon, the structure failed, unleashing a torrent that destroyed several ranches and much of the town of Rexburg, Idaho, downstream. Remarkably, only eleven people died.

In the aftermath of the Teton Dam collapse and the Oroville earthquake, the Bureau of Reclamation was well on its way to becoming the most controversial federal agency since the CIA. Later that year Jimmy Carter was elected president, and he went to the White House with a list of expensive federal projects he had promised to kill. One of them was Auburn. He was eventually forced to soften his position, and by the end of his term Auburn still had a budget and was being redesigned for greater seismic safety. Design work continued into the eighties as the Bureau sought money to begin construction again. Meanwhile, the diversion tunnel, the cofferdam, and the big bridge lived a longer and stranger life in limbo than their designers could have ever imagined.

After moving the dead man's pickup, I walked to the rail of the pedestrian walkway and looked over the edge. Beneath me the bridge's green steel trusses were in shadow. A car crossed the span behind me, and the clanks of the expansion joints startled a flock of pigeons from their roost underneath it. They veered out over the canyon as one, wheeled, and disappeared back under the bridge.

The body lay face-down below, where the canyon side had been bulldozed to subsoil and rock when the bridge was built and still nothing grew. There was a scuff on the soil just uphill of the corpse where the man had hit and bounced. He had come to rest with one of his legs twisted underneath him. There was a rent in his blue chambray work shirt at the middle of his back through which his intestines had exploded. They were splayed out on the ground around him.

The coroner's deputies arrived to collect him. They hiked down the steep rock and soil to the dead man, opened a body bag, and went to work. One of them took photographs. Another, a young woman, put on her rubber gloves and began daintily picking up bits of the man's internal organs and putting them into a plastic bag. She stopped for a moment and went over behind a rock to retch. Then she returned to work.

A television minicam operator showed up in a white van and walked onto the bridge to where I stood.

"Can you tell me who he is?" he asked me.

"Not at this time, sir," I answered.

"When did it happen?"

"About forty-five minutes ago."

"Did he leave a note?"

"Not that we're aware of."

"Oh."

The cameraman steadied himself on the railing and started filming a slow pan from the bridge deck down over the edge to where the deputies were working.

"You're not going to show him in that condition, are you?" I asked him.

"No . . . but I guess they would somewhere. I deal with Channel Ninety and Forty-seven, and they are mostly pretty tasteful about this kind of stuff. But it's a salable commodity, just showing the death scene, and I'm a freelancer."

I looked back at the deputies and the body. A log truck passed, and the bridge clanked and vibrated with the weight of the bones of some forest up the Foresthill Divide, headed for the mill in Rocklin. From under the bridge, startled pigeons burst over the gorge again, their black-pearl wings catching the morning sun.

Below, the deputies gingerly lifted the corpse into the body bag. Before they zipped it up, the woman collecting the organs put the plastic bag she'd been carrying inside it.

At the end of the day I was doing paperwork at the kitchen table in the old mess hall. Bell's patrol truck rattled into the yard, and he shuffled into the ranger station with his shotgun over his shoulder.

"Fuck this place," he muttered as he passed me.

He walked over to the gun cabinet and unlocked it with a large ring of keys clipped to his gun belt. Then, as he always did, he held the shotgun in front of him and with lightning precision racked six rounds of buckshot through the chamber and out the ejection port. The shells tumbled neatly through the air and landed in a small cardboard box full of loose shells in the gun cabinet, from which we all loaded at the beginning of our shifts and into which we all unloaded at the end of them. None of us did it like this, however. They didn't teach this at the academy, and for good reason. But it was Bell's trademark. He never missed the box, and although each shell passed the firing pin on its way through the chamber, he retired after twenty years without ever having blown a hole in the ceiling.

Bell gently stood the 12-gauge up next to the others and relocked the cabinet. A Vietnam-era veteran and expert shotgunner, he had a lot more respect for a Remington or a hunting dog than for a patrol truck or the government that owned it. On his way back past me, he glanced down at the report I was filling out in pencil.

"Your guy from the big bridge?" he asked.

"Yup," I answered, and sighed.

"A mess?" he inquired, walking over to his locker.

"The usual," I responded without looking up.

"Fuck this place," he said again. He opened his locker, took off his gun belt and uniform, put them away, and left to go home.

Before that, when I'd finished at the bridge, I'd gotten back into my Jeep and run up Foresthill Road. I turned around at Lake Clementine Road and drove back down to the bridge, more slowly than usual, with the window open. I didn't do it because this investigation required it. This one was simple: witnesses see man jump; no associates present; man very DOA; end of case. Instead, I did it because I always did. For a while now it had been my habit to construct a mental approximation of the events leading up to the matter I was investigating.

At the tip of the Foresthill Divide just before the dead man descended the last grade onto the bridge, there is a place where the new road to the bridge had been blasted through a hill of greenstone. Since the cuts were exposed in the seventies, tufts of apricot monkey flower, *Mimulus bifidus,* had grown all over them, in pockets of soil carried down from above by rain and gravity. In the late spring these incredibly tough plants were covered with thousands of azalealike yellow-orange blooms, and in recent years this unlikely spot had become one of the best places on the Bureau's land to see them.

Although thousands of people drove that road, I had never seen anyone stop and admire these gardens, and I doubted that the dead man did either. We are all so caught up in the struggles we get into on the way to the lives we dream of, and the dead man was probably just a little farther down that road than the rest of us. Maybe he had lost a good woman, a good job, or a good friend, or maybe he'd never had them. Or maybe it was bad chemicals, of internal or external origin, that pushed him over the edge.

But maybe he was just suffering from the same regret we've all known at one time or another, when life hasn't lived up to our expectations. Only his was worse, and perhaps his life lacked the sweet little mitigations that get most of us through our days: bandy-legged fawns on the lawn, a soft song you hum looking out on a parking lot with a cigarette in your hand, peach-colored flowers against gray-green rock, the company of friends, children, and animals, and the terse exclamations of your fellows, which let you know you are not the only one who suffers. Everything suffers. Everything has joy. In purgatory you still have a chance; the final judgment on you and everything else has not yet been rendered. So if people are doing something wrong, refuse to cooperate; if the music's too sad, for God's sake change the station or turn the radio off. Stop before the bridge. Get out. Walk down the road. Sniff the air, and if it smells good, breathe deep.

I idled down the road to the bridge. Here he got out and stepped over the kneewall between the roadway and the pedestrian walkway, then up onto the railing. For a brief instant he balanced between life and death. Far below him, the river was a set of whispering curves, its rapids seemingly motionless at this distance, like white paint on green glass. There were bright tufts of willow along its banks, and then the rocks and pines of the canyon walls higher up. But then it was too late because he'd already stepped into the lake of air, and there was the irrevocable quickness with which the wind increased in his ears and the battered earth came up to embrace him.

7 / A NATURAL DEATH

THERE IS A SORT of memory that does not refer to a particular day, yet it is not without precision, and accumulates from just being in a place for a period of years. Each time the American River floods big and brown with snowmelt and rain, I remember better the way huge drift logs turn ponderous circles in eddies, and where the river is carrying away land at the outside of turns, and where it builds beaches at the inside of them. Later, on warm spring days after the rains are past, I remember how little pink trumpets of bilobed clarkia and yellow daisies of eriophyllum float, as if mounted on some transparent medium, a certain number of inches, according to their species, above the steep hillsides; and how for 10 or 12 feet above that colorful surface there is a layer of air that hums and sparkles in the sun, composed substantially of insects seeking nectar.

I know where a tiny patch of a St. John's wort, called gold-wire for the shine of its filamentous stamens, grows tucked up under the chemise brush at the top of a red clay bank on a turn in the old Doc Gordon Road above Lake Clementine; it took me ten years to find it. Sometimes in summer one of the thunderstorms that boil up against the Sierra Nevada every afternoon reaches out as far as the foothills to the west, and a sweet damp smell rises from the dust just before the first drops of rain. The novelty of rain is one of the few things I liked about hot summers in the canyons, a season I mostly detested when I worked as a ranger in them. To be fair, however, the things I disliked about that time — the merciless sun that old forests would have shaded me from; the dust on my face, my uniform, and rescue equipment; the spiny star thistle that gets to flesh through thick jeans, wild oats that lodged in my socks, and the other disagreeable European annuals that overwhelmed the perennial meadows of the low Sierra — I eventually came to see as the marks of 140 years of bad treatment of this land. So over time I learned to forgive this place for its bad manners and prickliness, for these are the inevitable outcomes of servitude, in land as in people.

Aside from memorizing these natural phenomena that repeat themselves annually until the idle gaze comprehends them, I was content to let hours of work steal by without straining to save the details for posterity. I was a poor keeper of our required patrol logs. The quieter days were as seamless and unaccountable as water slipping by in the river, until time was apprehended by the duty to record something, such as the report, late in the day on April 23, 1994, that a woman named Barbara Schoener was missing up the Middle Fork.

The only thing I recall about that day before the call came in is an observation I made of the weather. At midmorning I steered my green Jeep into the entrance of the gravel road up the Middle Fork to the old limestone quarry. Turning off the engine, I looked east up the canyon where Barbara Schoener was at that moment, although I didn't know it. The sky was deep blue around harmless-looking puffy white clouds, the air was clear and cool, and the sun warmed my left elbow, out the open window. The riffles in the river whispered and sparkled in the eastern light.

Weather will be a deciding factor in any search, in the survival of the lost or injured, or if there is nothing left to do for them, in the difficulty and discomfort of recovering their remains. In this case, morning made a false promise. By nightfall the clouds gathered into a dark sheet and set upon the searchers, soaking them to the skin with a cold, steady rain.

What you do to investigate a death is a little like being a theater director. Unavoidably detained on the way to rehearsals, you arrive to find the final scene already played out and the actors and props spread around the stage in disordered repose. In your mind's eye, you send them back to their starting positions, marked in the theater with pieces of tape on the stage and in the woods by footprints, the victim's personal effects, and an unclaimed automobile at the trailhead. Then you set them in motion on the stage of your imagination, over and over, until you get it right. Later, when the report is written and the usefulness of thinking about it is long over, it's hard to forget this omniscient vision you've made of the victim's fated progress toward a bad end you know about, and she doesn't.

So it is that I see Barbara Schoener driving north from her home in Placerville. California poppies unfurl their glossy orange petals in the morning light between clumps of blue lupine along State Route 49, two lanes of winding asphalt connecting the string of little white-painted wood and red brick Gold Rush mining towns down the front of the Sierra. After half an hour she comes to the town of Cool, a county fire station and a group of plywood false fronts like a western movie set placed in an expanse of rolling pasture punctuated by stately blue oaks. She turns east onto State 193 at the only intersection in town, past the dirt turnout where scruffy men from the hills sell firewood out of beat-up trucks, advertising their loads with spray-painted signs on scraps of plywood.

Just east of there, Barbara Schoener passes the main gate of a residential development along the south rim of the Middle Fork canyon, expectantly named Auburn Lake Trails. Auburn Lake Trails is one of those gated communities that have turned old cattle ranches into recreational landscapes, with remnants of barbed-wire fences on split cedar posts going to rust and rot between big plywood houses on an aimless network of roads.

There are two more gates into Auburn Lake Trails in the next few miles east on 193, electric ones that can be opened only by magnetic security cards the residents carry. Barbara Schoener parks her car outside the second of these, across a perfectly paved road from the development's water treatment plant.

The woman who gets out of the car is forty years old, athletic, the mother of two children, with shoulder-length reddish brown hair. She wears a pair of blue nylon shorts, a cranberry sleeveless T-shirt, running shoes, a hat, and cotton gloves against the morning chill. She locks the car and puts the key in a little pouch attached to one of her shoes. Carrying an apple and a water bottle, she leaves the road, running down the trail into the neighboring state park.

At first she follows an old dirt road, grown over on either side by Scotch broom and narrowed to a single track. Horses and rain have worn a rut into the center of the remaining path; she places her feet with care. The road descends quickly into a Douglas fir forest, so that only a few feet from her car she is quite alone. Then the trail abandons the road, traversing the canyon side on contour, in and out of the folds of creeks. The trail emerges from the forest onto an open ridge. Far below, the river is spread out in a slow bend, silver against its gray gravel bed. She pauses to look and takes a bite of her apple, breathing deep of the air in which something bright — dust, a bit of pollen — catches the light out over the void. Ahead, entering the forest again, the path bends left into the manzanita.

At five o'clock in the evening, as I drove north on Highway 49 toward the ranger station to go home, the radio dispatcher called me and sent me back across the river into El Dorado County to meet with sheriff's deputies about a search in progress.

When I arrived, the missing woman's sedan was cordoned off with yellow crime-scene ribbon. Sheriff's search and rescue volunteers in orange shirts hustled around a mobile communications van. The wind was picking up. I got my jacket out of the back of the Jeep and shook hands with the officer in charge.

He said that when Barbara Schoener had failed to return as expected from a run, her husband had reported the matter to the sheriff. Her husband knew that she liked to run on this trail, and her car was soon found at the trailhead. She was probably equipped only with light clothing. The deputy and I agreed that I would drive up Quarry Road at the bottom of the canyon. There was a chance I would find her down along the river; when people get lost, they often head downhill until they get to something they can't cross.

It was dusk by the time I got back down to the rusty gate into Quarry Road. I let myself in and idled slowly east with the river on my left, watching the road shoulder on my right — we say "cutting it for sign" — for the lost woman's footprints. It started to drizzle, and I turned on the wipers. About two miles farther on, at Brown's Bar, the road became narrow and bad. The tires began to slip and throw bits of red clay up onto the hood. It grew dark.

This was the reassuringly familiar landscape of my nights — the interior of a Jeep, an exoskeleton of green humming steel, where I was surrounded by heated air and safe from most things, animals and weather, and, compared to a foot traveler, freed from the tyranny of distance. All businesslike: the tan upholstery of my seat, the lower right of its back torn from the constant abrasion of my pistol grips; the flashlight wedged between my right thigh and the radio console between the seats; and from the radio the cheerful blinking lights and a low chorus of calm voices from the rural counties around me, the men and women — police officers, rangers, paramedics, firefighters, pilots of medical evacuation helicopters — who come and go all night cleaning up scenes of chaos and imposing upon them the appearance of order that society requires in order to sleep well.

I turned on the rotating emergency lights on the Jeep's roof, so that if Barbara Schoener could move and was somewhere above me, she would see me coming from a long way off and have time to get to the road. Spokes of red and blue light circled around me, across the slopes of the canyon and the raindrops. I turned on all the spotlights, training them in different directions so I could watch for her, as I imagined it, waving urgently in the darkness. I reached down, punched up the loudspeaker, and pulled the mike off the dashboard.

"If you can walk, come down to my lights here on the road! Come to my lights!" I called over and over.

As I did so, it dawned on me that Barbara Schoener was gone. It was just a feeling, after all those years of searches, that I was talking to myself, that there was no one to hear me when I called to her. But still I called again and again, sending my amplified voice washing out over the cold boiling surface of the river, surrounding the dark trees and thickets of manzanita, filling up secret hollows.

Beyond Maine Bar, Quarry Road becomes two Jeep tracks across the sand and round gray cobbles of the gravel bars. I stopped to put the Jeep in low range and got out into the rain to listen and have a look around. Across the side of the canyon, several hundred feet above me, I saw the twinkling flashlights of other searchers through the mist of rain. I got back in the Jeep and bounced and scraped upstream across the boulders to the end of the road, watching the pools of my spotlights move across the feathery limbs of fir trees up the side of the canyon and straining to listen, through the roar of the rapids and the whine of the gearbox, for a cry in the darkness.

When MacGaff, O'Leary, Finch, and their fellow park rangers arrived in the American River canyons in January 1977, a team of park planners from State Parks' Sacramento headquarters was already there; their assignment, to prepare a plan for the development of recreational facilities on the Bureau's land. Finished in 1978, the General Plan for Auburn State Recreation Area featured a visitors' center, a lakeside snack bar, campgrounds for boaters, and a huge boat-launching ramp. The launch ramp was actually constructed, a steep cut in the canyon wall the size of a freeway that ended abruptly in thin air, hundreds of feet above the river. But the rest of the park's facilities were on hold until the dam's seismic problems were worked out. They wouldn't have done the rangers much good anyway, because like the launch ramp, these facilities were all about a reservoir that didn't exist.

With nothing to guide them, as time went by, the rangers managed the place according to the manifold desires of its most vocal users. Chief among these were the well-heeled residents of mini-ranches and horse properties in the foothills, where an annual hundred-mile endurance horse race over the crest of the Sierra had long been, for those on the social register, the equivalent of the symphony and opera in San Francisco. The Western States Trail on which this race was held ran right through Auburn State Recreation area, and the horse people soon saw the reservoir site as their private domain. Since State Parks lacked a clear vision for the place, the planning of a bewildering array of bridle trails through previously impenetrable (to human beings, anyway) thickets along the canyon walls was mostly done on the spot with pickup truck loads of shovels and mattocks delivered to volunteer equestrian trail crews by field staff — when, indeed, a ranger was even involved.

By the time Barbara Schoener was reported missing, no complete map of these trails existed in the hands of rangers or anyone else. In the search we used a sketch map that someone from the stables at Auburn Lake Trails had drawn. It showed only a portion of the south wall of the Middle Fork, but the maze of paths on it looked like a plate of spaghetti tipped by a careless elbow onto a restaurant floor. Some searchers got lost immediately, and the two of them I found wandering in the rain became the only people I would save that night.

However, if uncertainty about the dam's changing fortunes created in State Parks offices an atmosphere lacking in the sense of permanence that people who manage public lands ought to feel, the land in question lacked the capacity to equivocate. In the two and a half decades since the Bureau began acquiring it from its previous owners, it had begun to go seriously wild again. And no one at State Parks or the Bureau was studying this process.

Meanwhile, by the 1980s a long-awaited economic boom had taken hold in the foothills, based on residential construction, service industry, and the computer industry's arrival in western Placer County. Now the postwar land rush from cities into the dense grids of suburbs close to Sacramento leapfrogged outward into dispersed, low-density subdivisions of old ranches in the hills, where new residents could establish a relationship with nature, of the nice-view-from-the-deck kind.

In 1970 there were just over 20 million people in California. The population of El Dorado County, where Barbara Schoener lived, was around 44,000, which wasn't much, considering that at 1,711 square miles, Placer County's southern neighbor was just a little smaller than Delaware. By 1994 the state's population had grown to 31.4 million, just over one and a half times what it was in 1970. In the same period El Dorado County's more than tripled, to 146,400 souls. Most of them settled in the low-elevation western hills, where they would be spared the serious snow shoveling common in the high country and the commute to jobs in greater Sacramento was reasonable. And so, from the late eighties on, these counties along the foothill front of the Sierra Nevada were among the fastest-growing in the state.

Today, if you drive up into these foothills and allow yourself to wander, you will end up on dusty roads off other unmarked roads, which are in turn off other roads. At the end of each of them sits a relatively new house with no economic relationship, as a ranch house or a miner's cabin would have had, to the land around it. Everything that gets up there, from the next quart of milk to the next stick of lumber for a fence, arrives in an automobile, a pickup truck, or a sport-utility vehicle. It is a way of life unprecedented in history, and one so freshly arrived in these hills that some of its attractiveness may be residual from the activities of previous occupants — such as the vigilant extermination of predators carried out by cattlemen and by government predator-control hunters in their stead.

On the morning of April 24, the second day of the search, four young men, long-distance runners and acquaintances of Barbara Schoener, went out to look for her. At about 7:15 A.M. they found her water bottle along Ball Bearing Trail. Nearby they saw signs of a struggle in the steep duff below the trail. They followed these marks just far enough to see Barbara Schoener's feet sticking out of a pile of sticks and forest litter farther into the draw. They ran back to the trailhead and reported that they had found her, dead.

I made arrangements with one of the sheriff-coroner's deputies to meet for a death scene investigation later in the day, upon the arrival of a forensics specialist from the Department of Justice. It was either a wildlife killing or a homicide, he said. Hard to say right now.

Around four in the afternoon I drove through Auburn Lake Trails to the edge of the canyon. There I met the sheriff's deputies and the forensic technician, a cheerful woman named Faye wearing a blue jumpsuit. Hands were shaken, introductions made, and outcomes of other recent investigations asked about as we loaded cameras, equipment, and body bag into backpacks.

We started up the trail at the opposite end from where Barbara Schoener had entered. We were at most about half a mile from the nearest house, a conventional tract home with a two-car garage on a neat asphalt cul-de-sac, sitting in the middle of the grass and trees as if it had dropped from the sky. The switchbacks we walked up in the first one hundred feet of the trail were brand-new work. Being killed by an animal on this trail would have been unlikely thirty years before. The trail hadn't existed then, and neither had the road to the trailhead, nor the house.

We came to a place where the path traversed a hillside, falling off steeply to the left in the shade of oaks and firs. Some horses with orange search and rescue equipment hanging from their saddles stood in the trail, tied up to trees. Just beyond them, we ducked under a barrier of police crime-scene ribbon. Three sheriff's search and rescue volunteers were setting up fixed ropes down the steep hillside to assist us in getting the body out. They spoke to one another in low, funereal tones. I nodded a greeting to each of them as we passed, and they nodded back grimly.

About three hundred feet farther, the trail curved out from under the trees onto a rocky ridge surrounded by manzanita bushes that provided cover close to the path for anything that would have wanted it. One of the sheriff-coroner's deputies pointed out a divot of moss loosened from the bank uphill of the trail.

As we reconstructed it, the cougar had been sitting in the brush, maybe hungry, lying in wait for the next animal to come along the path. That happened to be a runner, a woman. The animal sprang down the bank, leaving the divot as it launched. It hit Barbara Schoener from above and behind. She staggered into the soft duff downhill of the trail, where her feet left two unmistakably deep impressions; she was heavy with the weight of both of them, struggling to remain standing. She went down against a fallen fir sapling that lay across the slope below the trail. There had been a struggle: branches were broken off the dead tree, and a dark stain on the soil smelled a way you don't forget. She stood up again and staggered downhill over the tree. There were a couple of more footprints in the duff. Just below and to the left of these, at the base of a Douglas fir sapling, I found one of Barbara Schoener's cotton gloves soaked red with blood and a Red Delicious apple with her dainty bites around its circumference.

From where these things lay, scuff marks led down to the bottom of the steeper part of the slope below the trail, maybe one hundred feet. There the dragging started, leaving a furrow in the ferns just as wide as a small woman's body, which continued another one hundred feet to where we found her.

Once, years before, as a boy hiking alone in the forests near my home, I had stumbled on a mountain lion kill, a deer, dragged into a cool canyon and covered with sticks so the cat could come back later to eat more. Nothing was scattered around; there was a kind of fastidiousness to it. I recognized it immediately when I saw Barbara Schoener.

The legislative history of cougars in California reflects the change in attitudes toward predators with the growth of an environmental ethic in the 1960s and 1970s, and the changing demographics of the state's electorate from rural-agricultural to primarily urban-suburban. Increased protection for mountain lions would probably not have occurred had the legislature continued to reflect the wishes of any sizable constituency who had their next mortgage payment tied up in a flock of stupid and defenseless sheep standing around at night in a remote mountain meadow.

Up until 1963, if you saw a lion, you could shoot it and collect the bounty from the state government: $50 for a female and $60 for a male. This reward was sometimes enhanced by counties.

There may have been only two thousand lions left in the state by midcentury. After 1963 the bounty was removed and the cougar was classified as a nongame mammal. In 1969 California reclassified the mountain lion as a game mammal, and for two years permits were issued to hunt it for sport. In 1972, as younger wildlife managers brought an appreciation for the role of predators in healthy ecosystems to their work and as concern grew for the shrinking populations of cougars, lions were entirely protected from hunting, except as necessary to protect lives and property. The shape of this protection has not changed appreciably to the present day. It has always been possible to get a depredation permit to kill one or several lions if they threaten people or livestock. Troublesome lions are generally exterminated by the ranchers who apply for the permits, by trappers from a branch of the U.S. Department of Agriculture calling itself, somewhat euphemistically, Wildlife Services, or by wardens from the state's Department of Fish and Game.

In 1990 California's voters passed a law known as Proposition 117, confirming their preference that the cougar not be hunted. The new law also created a habitat conservation fund for the purchase of wildland areas that cougars inhabited. Since much, maybe most, of California is cougar habitat, this provided acquisition money for parks and reserves that in practice protect all kinds of other wildlife that happen to inhabit them.

But critics of the cougar's protected status say that under the current regime, the state Department of Fish and Game has not been allowed to manage the overall population growth of cougars by opening a hunting season. They say the cougar population is expanding out of control. They point out — and most biologists agree with this point — that the pattern of habitat utilization by cougars involves the ejection of younger animals from more desirable remote country already occupied by dominant older animals into marginal areas along the suburban edges of wildlands. This, say the lion's critics, will increasingly bring young cougars into contact with people in the suburbs, and before long they will be experimenting with stealing children from their bicycles. That's hysteria, say the cougar's supporters.

Our retribution for the death of Barbara Schoener was swift. When I returned at seven in the evening from picking up the body, I notified the Department of Fish and Game's dispatch office that I believed I had just investigated the first killing of a human being by a mountain lion in the state in the twentieth century. Fish and Game notified the U.S. Department of Agriculture trappers, and the following morning I returned to the scene with two trappers and three Fish and Game wardens to begin the hunt.

Lions are hunted either by staking out a captive farm animal as bait and hiding nearby or by running them down with specially trained dogs. In the latter process, a houndsman will drive a rural road with the lead dog standing on top of his specially built truck until the dog, crossing a scent on the wind, gives voice. Then the trapper lets the rest of the dogs out of their cages in the back of the truck and puts them on the chase. A single dog is no match for a lion, but lions are scared by a pack. A lion will climb to a high spot, a cliff or a tree, and the dogs will keep the animal there and howl at it until the hunter catches up. Then it's like killing fish in a barrel, if you have a rifle or pistol.

That's basically how it was done this time. After eight days of tracking and dog work, the Department of Agriculture's hunters picked up the lion's scent when it came back to the scene of the kill, most likely to feed again. The chase was short, and the cougar was treed and shot a half-mile away, on the other side of Maine Bar Creek.

The dead lion was an eighty-three-pound female. Barbara Schoener outweighed her by about twenty pounds. The lion's udders were full, which meant she had a cub. Over the next couple of days the trappers went back out and found a kitten, which was displayed for the news cameras and then turned over to a zoo.

A few weeks later I sat in a lecture theater at the University of California Medical School at Davis, among newspaper and television reporters, state officials, and representatives from animal rights organizations. A procession of experts in forensic fields took the stage and described how they had identified the dead cougar as the one that had killed Barbara Schoener. A forensic odontologist had matched the animal's teeth with impressions in Barbara Schoener's crushed skull. Experts in DNA typing had swabbed the folds of skin into which the animal's claws retract (as a housecat's do) and located human DNA — not just any human DNA but of the same type as Barbara Schoener's. They had killed the right lion.

It did not stop there.

During that same year, at another California state park hundreds of miles to the south, just north of the Mexican border, there had been a series of disturbing encounters and close calls between lions and hikers. One ought to be cautious about drawing conclusions, yet there are some resemblances between the places: Like Auburn, Cuyamaca Rancho State Park had for many years been working land, a cattle ranch, before it was deeded to the state park system. And like El Dorado, the brushy hills of San Diego County had seen a massive invasion by housing tracts.

At Cuyamaca in December 1994, within eight months of California's first such modern-day fatality, fifty-six-year-old Iris Kenna was dragged off a fire road and mauled to death while hiking alone near a popular campground. It would be nearly a decade before another death, that of a thirty-five-year-old mountain biker, Mark

Reynolds, who was attacked and killed while riding in the hills of Orange County in January 2004. At this writing, there have been eleven incidents in which lions attacked people in California since 1890. Three involved two victims each. Eight have occurred since 1986. Four people died from their injuries, and in 1909 two more died from rabies they contracted after a lion attacked but did not kill them.

The overwhelming majority of the roughly 230,000 Californians who die each year succumb to disease. Those who die from physical trauma do so principally as a result of their own actions or those of other people, not the actions of animals. In 1994, the year in which two women in the state were killed by mountain lions, there were 3,821 homicides and 4,212 traffic fatalities in California.

Nevertheless, after the 1994 killings, an angry "eye for an eye" sentiment prevailed among conservatives in the state legislature. Several bills to reinstate sport hunting of lions with dogs were introduced, supported and sponsored by the huntsmen's clubs. An initiative statute was prepared for the March 1996 primary election to repeal the protection mountain lions had enjoyed from hunting and to assign Fish and Game to manage and control the cougar population. The voters roundly defeated it. Urban and suburban people in California like their wildlife. Most of them have never even seen a lion, and many would like to see one, under the right circumstances.

I have seen only three cougars in my life, twenty-one years of which I spent working as a ranger in lion habitat. One of them ran across Foresthill Road in front of my Jeep about two miles as the crow flies from where Barbara Schoener was killed. I patrolled these roads in Jeeps, the trails on foot, and the rivers in boats for eight more years without seeing another. But I'd wager they've seen me, often. Remaining concealed is what mountain lions do for a living: They hunt as housecats do, hiding or quietly stalking until they pounce.

How many lions are there in California? No one really knows Because they are hard to see, cougars are hard to count. And be cause they are hard to count, it would be difficult to manage their populations in any precise way. In 1988, the last official study of the state's lion population resulted in an estimate of 5,100. Official figures show that well over 2,000 of them have been killed under permit in the state since 1972, and the pace is quickening: More cougars were killed in the last decade than in the preceding two decades, and by early 2004, one hundred more had been shot in the first four years of the new millennium than had been killed in all of the 1980s. There is no assurance, say experts, that lion populations can sustain such losses in the long run. In the meantime, those that survive exist in the spaces between thirty-six million people and countless domestic animals who are actively invading the wilds.

When I came to the American River, I thought a ranger's job was to save something, or someone. Sometimes it is, when you hear about a bad situation early enough to stop it before it happens. But so often — as in the case of Ricky Marks and Mary Murphy — the whole story unfolds one step ahead of you. Or it's all over and done with before you even hear about it, as it was in the matter of Barbara Schoener. Then all you can do is to try to memorize the details and give a good account of them in your report. As time went on, it became clear to me that this was an important part of my job, too. A ranger is privileged to be intimate with things few other people spend much time with, and your job is to witness and remember.

What my memory had distilled from eight years of witnessing in the American River canyons by the time of Barbara Schoener's death was a glimpse of the general direction of things there: the return of a cool ponderosa pine and Douglas fir forest, pushing up through live oaks and black oaks; the profusion of wildflowers in the meadows after cattle were removed; and the growing frequency of sightings and tracks of mountain lions and black bears. After a century and a half of condemnation to usefulness there was a great longing back toward wildness in these canyons, and they had begun to go that way with an energy like continental drift, like roots heaving pavement. It was desire; it was the force behind everything that happens without human permission or design. It is present in the heartbeats of tiny birds who roost in trees on nights when we would quickly perish from exposure, if not for our houses and warm clothes. When this energy brings the missing parts back to a place, it can be uneven and unpredictable or, as it was for Barbara Schoener, even dangerous.

I will never forget how she looked, surrounded by the way things are at that time of the spring on a north-facing slope: young Douglas firs, green ferns, and moss in the dappled light.

At five o'clock in the evening on April 24, as the low sun turned the tops of the conifers orange, four of us walked down into the forest below Ball Bearing Trail and knelt in the ferns around the neat mound of duff and sticks that covered the body, except for the top of the scalp and the neatly tied running shoes. The evidence technician and I began to remove the pile of twigs one at a time, inspecting them for animal or human hairs, which we collected by touching the twigs with pieces of adhesive tape and then sticking the tape to white evidence cards. The two sheriff-coroner's deputies presided, taking notes and labeling, and packaging the evidence in brown paper shopping bags.

The little glade seemed strangely peaceful.

When we finished, for a moment the glistening internal wilds of spine, ribs, and intercostal muscles looked like food, like the inside of some deer. We gently rolled her over. Her face was gone. From below us through the deepening shadow of the forest, the roar of the rapids along the Middle Fork rose and fell on a breeze. That sound is behind everything I remember in those canyons, like the sound, or a name I know but cannot pronounce, of some larger turning of things into other things. We photographed her, and then we put her in a body bag and bore her back up the hill.

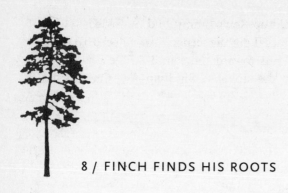

8 / FINCH FINDS HIS ROOTS

SOMEONE HAD BEEN STEALING TREES from the prettiest groves of black oaks in our forty-two thousand acres, at a place on the west rim of the North Fork called Big John Hill. Ranger Ron O'Leary had been checking the area for weeks, but so far he hadn't turned up a lead — just, each time, more stumps.

One April morning a neighbor along our boundary heard a chainsaw running and called the Sheriff's Department. Knowing that Big John Hill was part of Auburn State Recreation Area, the deputy who took the call asked his dispatcher to have a ranger handle it. I heard their conversation on my scanner and started rolling that way before our dispatcher called me. On the way I radioed O'Leary, up at Mineral Bar.

O'Leary and I rendezvoused near the village of Weimar along Interstate 80 — a grocery store, a gun shop, a tiny post office, and some rough little dwellings — and from there I followed him east along the wooded toplands toward the canyon of the North Fork. Dropping into Specimen Gulch, the road turned to gravel. At our boundary it devolved into a few pairs of red-clay ruts wandering off in various directions into the pines. We stopped and got out to listen. The cool air was pungent with bear clover bruised by our tires. To the east toward the canyon rim, we heard the complaint of a chainsaw. We locked up Ron's Jimmy, and he folded himself in behind my shotgun rack. I put the Jeep in four-wheel drive, and we followed one set of ruts toward the sound of the saw. The ruts climbed steeply onto the groves of Big John Hill. On all sides of us the charcoal-gray trunks of black oaks split into symmetrical, wineglass-shaped bowers of branches supporting a ceiling of spring leaves sixty to seventy-five feet overhead. The leaves hadn't sunburned to the darker color they would be by midsummer, and the morning sun filtering through them bathed everything beneath in luminous pale green.

The track we followed forked often, and at each fork I stopped to listen for the saw, then steered toward it. Soon we saw stumps and tangled piles of limbs on either side. By now the trail was nothing more than two wheel marks of crushed vegetation slaloming through the trees. A stick popped under one of our tires. I stopped to listen out the open window for the reassurance that the saw was still running. It was.

Toward the hillcrest, our way was blocked. The tree poachers had felled an oak across the trail to keep prying eyes away from their larceny; no doubt they planned to cut their way out when they left. The dense forest on either side afforded no other way through. We didn't have a saw, but the Jeep was equipped with an electric winch. I got out, hit the gear release on the winch, grabbed the hook at the end of the cable, and, leaning into it, paid out the cable toward the fallen tree. By the time I'd made it fast, Ron had the winch control in his hand. He hit the switch, the cable went taut, and the whine of the electric motor dropped half an octave as it took up the load. Rustling and snapping branches, the tree began to inch toward us. A hinge of bent and splintered wood still connecting trunk to stump suddenly gave way with a loud snap. The saw, no more than two hundred yards away now, stopped abruptly. Ron took his thumb off the winch control and we stood stock-still, hardly daring to breathe. The rapping of a woodpecker echoed through the forest.

The saw started again. Ron hit the winch control and the broken stubs of branches on the fallen tree dragged dark grooves in the forest floor. The saw stopped again. Ron stopped the winch, and again we held our breath. The saw started, and again so did the slow progress of our tree. Eventually we had enough room to sneak the Jeep between the tree and its stump. We got in, closed the doors gently, and continued. O'Leary removed my camera from its case on the floor.

Around another bend in the trail, suddenly they were right in front of us: two men and two pickup trucks, both partially loaded with oak firewood. I stomped on the accelerator to close the remaining distance. To my right the camera flashed off the inside of the windshield as Ron caught them in the act.

The photograph would show one of them — a stocky man with a drooping mustache, his muscular arms below his T-shirt covered in bluish tattoos — just looking up to see us. The other — huge and lanky, with long, stringy blond hair, wearing a red flannel shirt and jeans — would be captured still intent on his saw. A whitish blur of wood chips was frozen in midair, cascading down his pant leg.

We bailed out and began walking toward them. The first man tapped the sawyer on his shoulder. The other looked up. The tattooed one nodded darkly in our direction. The big man — his driver's license would show he was six feet seven — hit the kill switch on the saw and uncoiled himself upward to gape at us.

"You're under arrest for cutting park trees!" I called out as I walked toward them. "Keep your hands where I can see them."

I moved in, patted them down, removed their knives — both men carried them — handcuffed each in turn, and sat them behind the prisoner cage in the Jeep's back seat. O'Leary stood back, watching carefully, which one ranger always does so the one preoccupied with searching and handcuffing doesn't get taken by surprise. The two men said nothing. When I closed the door on them it was just after eleven A.M. The day was growing warm, and the air was full of the rich scent of spring — oak leaves, deer brush flowering, bear clover, and moist soil. O'Leary was on the radio, requesting two tow trucks.

To most Americans a ranger is a nostalgic figure, living a simple outdoor life reminiscent of that in the nineteenth-century American frontier. Surrounded by herds of elk and the world's tallest trees at California's Redwood National and State Parks, she's a jack-of-all-trades who can splint a broken bone and replace a busted fan belt on her truck with equal facility. At Sequoia and Kings Canyon in the high Sierra, he might ride up to you on horseback, wearing the flat-brimmed campaign hat that is the ranger's most recognized symbol. At Glacier Bay in Alaska, she might paddle up in a sea kayak to explain the habits of grizzly bears, in whose proximity she beds down each summer night unprotected but for a tent and a can of pepper spray. In Glacier National Park in Montana, he might appear out of an early June sleet storm to orient you on an indistinct trail across a high pass. Answering visitors' questions in a cliff dwelling at Mesa Verde in Colorado, her knowledge of Indian culture might be deeper than that gleaned from book-learning, for your ranger could well be Navajo or Hopi herself.

All of that feels very Old West-y, but in fact the ranger is a distinctly modern figure, who didn't appear until the historical moment — between 1850 and 1900 — when it became possible to imagine nature as no longer an adversary but a conquered and cornered thing in need of preservation. That the first rangers were contemporaneous with the emergence of the Arts and Crafts Movement in the field of design is no accident. By the second half of the nineteenth century, the great cities and their steam-powered factories had become smoky, clamoring horrors and mass-produced factory goods were on their way to eradicating the mark of human hands on the things people used, wore, and lived in. Some urban people — the upper classes, anyway — were suddenly filled with a yearning for rusticism: fabrics, furniture, and architecture with the marks of a craftsperson's hand, and the clear air, uncrowded vistas, and birdsong that were increasingly obscured by coal smoke, miles of brick, and the deafening cacophony of the streets.

The Arts and Crafts Movement is said to have begun in England, but national and state parks were an American invention. Parks were meant to evoke the presettlement landscape, and their advent was driven by an immediate nostalgia for the western frontier at the time of its closing. However, where the frontier was characterized by expansiveness and progressive subjugation, parks were created by a new and diametrically opposite force: the intentional refusal of progress and the encouragement of picturesque primitiveness.

Today at a park or wilderness boundary, all sorts of things are said no to: roads, off-road vehicles, loud music, fireworks, and firearms. The U.S. Forest Service has at times interpreted the Wilderness Act as prohibiting the agency's own use of chainsaws for trail maintenance, bulldozers for firefighting, and generators to light ranger cabins within designated wildernesses. An American company that produces washboards, hand washers, and mangles for the Amish (whose religion forbids the use of some modern contrivances) has among its other customers the National Park Service, which supplies these antique laundry implements to some of its rangers in remote backcountry cabins.

The creation myth of the national park idea is a scene oft related by historians: In September 1870, seated around a campfire, members of an expedition to survey the natural wonders of the Yellowstone Plateau talked about the area's future. Some of them felt Yellowstone's geysers, hot springs, and other curiosities ought to be leased to entrepreneurs for development. But one, an eastern lawyer by the name of Cornelius Hedges, is said to have maintained that, rather than fragment Yellowstone's marvels into private hands, they ought to be protected in a public park.

Returning to the East that winter, Hedges's fellow expedition member Nathaniel P. Langford was sufficiently taken with Yellowstone and Hedges's vision for it that he spent the winter lecturing and publishing on the subject. He was sponsored by Jay Cooke of the Northern Pacific Railroad, who saw a Yellowstone park's potential to boost passenger receipts as a tourist destination. The railroads would later be influential in pushing legislation for Yellowstone and other parks through Congress.

In the course of his speaking tour, Langford is said to have caught the attention of Ferdinand V. Hayden, director of the federal government's Geological and Geographical Survey of the Territories. The following summer Hayden organized another expedition to Yellowstone, taking with him the photographer William Henry Jackson and the landscape painter Thomas Moran. With their powerful images assisting the cause, Congress began deliberations on a park bill in December 1871. The act creating the park was signed into law the following March by President Ulysses S. Grant.

It was a paradoxical moment. While the pace of settlement and the indignities of modern life in cities already made it clear to people like Hedges and Hayden that the remaining slices of the unfenced West were endangered by and would soon be needed by their countrymen, the park's creation predated by four years the 1876 slaughter of Custer's Seventh Cavalry by twenty-five hundred Cheyenne and Sioux warriors in Montana Territory — a very frontierlike event. Still, by then the United States was linked coast to coast by steel rails and copper telegraph wire, and the thirty to seventy million bison that only recently had thundered across once endless seas of grass on the Great Plains were well on their way to extinction. Of the handful that survived in the United States by an 1889 census, the largest herd, of about two hundred, had taken refuge in Yellowstone, where they would ultimately be saved.

That, then, is the official version of the beginning of national parks.

But in fact the first national park wasn't one at all. It was a state park called Yosemite, and the first California state park ranger there predated Yellowstone by fifteen years. And I wouldn't have known this if it hadn't been for Finch's midlife crisis.

———

Our American River canyons were the inverse of Yellowstone: they were preserved only by stays of execution. Our work in them was frequently dangerous, and we rangers depended intimately upon one another for safety. For years we spent more of our waking lives with each other than with our wives. Yet we communicated our feelings to each other in only the most indirect ways — by casual inference, in pointed jokes, and with innuendo — and the thing most unsaid between us was the daily agony of risking our skins for nothing. In the face of this, to preserve our mental health, each of us learned to cultivate interests outside our jobs.

I had my writing, a wife, and eventually two children.

O'Leary had a wife, a little boy, and a beautiful home he'd built with his own hands. He and Bell were co-owners of a salmon boat they kept trailered in our truck shed. A picture of O'Leary at the helm, motoring across a coastal inlet against a fog bank with a full catch, hung on our office wall. He seldom beamed like that at work.

Bell was a hunter, passionate about pheasant and turkey season and his nervous, amber-eyed vizsla bird dogs. On summer evenings he played softball with a team sponsored by a local cabinet shop. He had a wife, a son, and a daughter who dreamed of becoming a ballerina.

Sherm Jeffries had his wife, two daughters, his church, his fly-fishing.

MacGaff had courted his wife in the mountains and each summer they'd go camping at the place where they met. He kept a garden and made his own beer. But more than anything, his accounts kept the ship of his life on an even keel, and he trimmed and balanced them the way a sailor trims sail. Each day he knew the exact balance of our park budget, the exact number of days until his retirement, and the precise amount accruing to him in his pension plan. He could quote to the nearest quarter-hour his vacation hours. He maintained himself in a state of robust good health not as an end in itself, we all suspected, but as a way to hoard a mountain of unused sick leave for which the department would have to pay him when he retired.

Below the dam's waterline each of us was left to work out his salvation in his own way, and one of MacGaff's ways was to keep an inventory of everything he might be able to salvage from the place before it went underwater. At one location where the Bureau had burned a former resident's house, MacGaff would note a tumbledown chimney. Later the bricks would disappear, and a few weeks after that they'd reappear in a decorative path and a little brick wall around a flowerbed at his home. Elsewhere in our canyons he'd come upon a pile of large planks, the remains of a former resident's barn. Sometime later they'd disappear. Sometime after that, raised beds in his garden where he grew strawberries would be surrounded by neat plank boxes. MacGaff kept notes on these sorts of resources in a notebook in the breast pocket of his uniform. It was legendary. One ranger claimed to have inspected it one night when MacGaff went home, leaving his uniform shirt draped over a chair. According to that ranger, it contained entries like "firewood: large fallen oak limb, Windy Point, approx ¼ cord."

And so it did not escape notice when MacGaff began taking a five-gallon plastic bucket with him when he drove away from our station each morning. Each evening when he returned, he'd carry the bucket — now obviously heavy — from his government Jimmy to his own pickup. The next morning the empty bucket went back in his Jimmy before he went on patrol. Eventually it was learned that MacGaff had graveled his whole driveway, one bucket at a time, from an abandoned gravel quarry five hundred feet beneath the dam's waterline.

Finch had his union work. In 1979 state park rangers were the lowest-paid law enforcement officers in California and, fed up with low wages and capricious discipline from desk-jockey superiors, Finch decided to organize. Gathering around him a cadre of dedicated colleagues — O'Leary was the first, and the union treasurer — Finch negotiated pay parity with other cops, recourse in the face of unfair discipline, bulletproof vests, better guns, and new patrol wagons before the old ones fell apart underneath us. Then, in a final flourish, he got us sixty dollars a month physical fitness pay if we could pass an annual exercise test. But eventually the state park rangers' union was gobbled up by a larger union of state employees and Finch resigned his presidency in disgust. Now he had only his job in our doomed canyons, and his fortieth birthday was coming up, too.

For the first time since I'd known him, Finch seemed sullen, preoccupied.

One day in the kitchen I asked him, "What's eating you, Dave?"

"I don't know," he answered. "The ol' midlife crisis, I guess."

The boundaries of parks and wildernesses are really just lines on a map. In practice they are permeable to air pollution, tree diseases, and the peregrinations of eagles and mountain lions, feral cats that hunt songbirds, and domestic dogs that chase deer. Most of all, park boundaries are permeable to human behavior, because people bring their problems with them when they come. Those who commit crimes in a park are generally the same who transgress against their fellow citizens elsewhere. And since the 1960s, with population and social problems growing outside their parks, rangers increasingly spend their time defending not trees and animals but the experience of their visitors — their peace and quiet and safety — from other visitors.

One summer evening Finch got dispatched to investigate shots fired in the campground at Upper Lake Clementine. O'Leary and I responded to back him up. Finch got there first. At the bottom of Upper Lake Clementine Road in those days, what we called a campground was nothing but some tracks in the sand through a jungle of willow and giant bamboolike arundo, ending on a beach where people set up their tents. We had a campground host living in a camper at the entrance to keep an eye on things.

When Finch got there the host, a man I will call Bob, told him that a particular group had been shooting a large-caliber revolver in the campground for over an hour. Bob gave Finch his detailed notes with descriptions of the people and their vehicles. Then, just as Finch made ready to leave, a brown pickup truck drove by, headed out of the campground.

"That's one of the trucks!" exclaimed Bob.

Finch gave chase, hit the siren, and stopped the truck on the road out of the canyon. He had no idea at that point where the gun was. Approaching cautiously, he found that the truck's single occupant was a woman. She looked nervous. Finch questioned her about the gunfire.

"What gunfire?" she asked. When someone says that about a .44 magnum going off repeatedly in a campground, you know you're close to the suspect.

But before Finch could go any further with his investigation, a man drove up on a motorcycle. He had an open beer clamped between his thighs. He demanded to know what Finch was doing with his wife. The man fit the description of one of the shooters. He was conspicuously drunk, so Finch detained him for drunk driving.

O'Leary and I arrived. We got a brief account from Finch of what had transpired so far. No sooner had Finch finished than another pickup truck drove toward us. It too matched one of the vehicles Bob had described, and when it pulled up alongside us we found its driver as drunk as the first two. O'Leary and I detained him as well. Like shooting, drunk driving is dangerous in a campground, where it's possible to back your car right over someone in a sleeping bag in the dark.

By now things were getting a little hard to keep track of. It was easier for three officers to carry out one arrest than it was for us to arrest three unpredictable drunks at the same time. Then, as O'Leary was walking the second of the drunks to Finch's Jeep, a fourth member of the party appeared on foot.

"What are you assholes doing? Leave him alone!" she slurred, making a beeline for O'Leary, as the latter attempted to pour his prisoner into the Jeep's back seat.

Finch stepped into her path. She shoved him. He stood fast. Behind him O'Leary was still struggling with his man, who wasn't following directions. I was over with the first woman Finch had stopped, trying to keep her out of the fray.

The other woman kept on shoving Finch and screaming, "You fucking assholes, you fucking assholes!"

"It's time for you to leave," Finch told her in a voice that was loud but surprisingly calm. "Walk away and stop interfering, or I'll have to arrest you," he said.

"Fuck you, you fuckers! Leave him alone!" the woman answered at the top of her lungs, pawing at Finch to get at O'Leary.

Finch had had enough.

"That's it," he said, grabbing one of her wrists as he reached for his handcuffs. But she wasn't going for it, and the fight was on. Now she was trying to bite him. O'Leary closed the door on his prisoner and turned to help Finch. I told the first woman to remain in her pickup and went to help Finch, too.

By now it had grown quite dark, and distracted by the scuffle, none of us saw the woman I'd left in the first pickup get out and sneak around the other side of Finch's Jeep, where she opened the back door we had mistakenly left unlocked and let both prisoners out. O'Leary and Finch and I were struggling to handcuff the screaming woman when we looked up and saw both men out of the car, staggering around in their handcuffs, yelling about police brutality. Having worked the second woman into her cuffs, O'Leary and I left her kicking at Finch and went to corral the men. Another struggle ensued. We dragged them back to the car.

Eventually we had all four drunks in the back seats of our vehicles. We were sticky with sweat and covered in dust. The combination made mud. We radioed for a brace of tow trucks. In the impound search we found the gun, a loaded .44 magnum revolver, under the driver's seat of the first pickup Finch had stopped. Why hadn't it been pulled in the melee? Just luck, we guessed.

After Finch departed from the union, his quest for diversions diversified. He became an avid collector of old ranger badges and uniform insignia.

One day I came into the office at the lower end of our compound and found him at his desk. Spread out in front of him and across a typing table to his left were old photographs and hand-tinted post-cards.

"What's all this? New hobby?" I asked him.

"I've been looking for old photos of rangers — the first rangers. I got these from a guy at a badge collector show."

"Who's this guy with the beard, standing in front of the tree?"

"That's Galen Clark, the first Guardian of Yosemite," answered Finch.

I picked up the photo.

"Pretty wooly-looking — a real frontiersman, with the beard, that rifle, and the mountain man costume. Early national park ranger, huh?" I asked.

"Nope, state — *state* park ranger. The first. His actual title was Guardian of Yosemite. Yosemite Valley was deeded by ol' Abraham Lincoln to the state of California as a public park in 1864."

"I thought the army took care of Yosemite."

"That's true," replied Finch, "but they didn't get there until 1891. Our guy Galen was there first. From what I can tell, he was the first ranger in the United States, and he was a state ranger, just like us."

The actual mechanics of defending Yellowstone and Yosemite National Parks did not emerge fully formed. In 1866, when Galen Clark was appointed by the California Legislature to be Guardian of Yosemite, he was one man living in the midst of a spread-out community of homesteaders, innkeepers, hunters, and sheep-herders making a living off the land he was supposed to be protect-ing. The lumber for the settlers' habitations and guest accommoda-tions, the grass and hay for their horses, and the wild and domestic meat and vegetables that graced their tables most often came from park territory. As a result, much, maybe most, of the energy ex-pended by Clark and the other first guardians went into controlling the depredations of their fellow residents.

Still, from the very beginning the pattern was familiar. Researching his roots, Finch traveled to Yosemite, where he learned the circumstances of Galen Clark's first known arrest. In 1870, Clark apprehended two men who had cut down a huge pine tree. He took them before a judge in Mariposa, where they were convicted and fined twenty dollars each. Finch also located a report by Clark's successor, James Hutchings, to the California Legislature of 1882.

"Here it is — listen to this," he told me one day from his desk. "'Sometimes we are visited by rough characters from the mountains who, when crazy with liquor, not only become nuisances, but sometimes endanger human life.' Sound familiar?" he asked.

"Some things don't change," I replied.

"Yup," he said, smiling. "Some things never change."

But Clark, Hutchings, and the other solitary guardians could never effectively patrol and protect the hundreds of square miles of the early parks. In 1875 it was reported that four thousand elk had been slaughtered by wildlife poachers in Yellowstone the previous winter, and five years later, an estimated ten thousand annual visitors were under no practical supervision in most of the park. They, their innkeepers, and their guides went around cutting down trees, shooting animals, and chipping souvenirs from the rock formations of Mammoth Hot Springs. In an attempt to remedy the situation, a local mountain man, Harry Yount, was appointed to guard Yellowstone. He resigned after only a year, complaining that the task was hopelessly large for just one man. A series of government investigations of conditions in national parks during the following decade resulted in scathing reports on the failure of civilian authorities to protect them properly. Some stronger force was needed, and into this void, in 1886 at Yellowstone and in 1891 at Yosemite and Sequoia National Parks, were sent detachments of U.S. Army cavalry, whose superiors had long expressed an interest in the job.

The first troops were under orders to evict squatters, capture fugitives, protect natural features and visitors from harm, and control depredations by innkeepers and guides. But their commanders intuitively invented the larger trade of park management in a rough form of what it has become today. They made maps and surveys of plants and wildlife. They constructed trails, roads, and headquarters. They stocked fish, fed herds of elk and buffalo through hard winters, and began closely regulating the activities of hoteliers and other park concessions.

And so, if our pedigree as rangers goes back to the never-uniformed Galen Clark and Harry Yount, we are more recognizably the descendants of these uniformed and intensely bureaucratic turn-of-the-century cavalrymen. From them came the flat-brimmed cavalry hat rangers still wear, and from them the olive-green uniforms, which had supplanted the army blues by the time the Park Service took over from the army during the First World War. This horse-soldier army, the historian Harvey Myerson has remarked, existed on rules and regulations; its lifeblood, the orderly flow of paper forms for every conceivable occasion through successive ranks for approval. Today California state park rangers have no fewer than 142 forms and I spent about a third of my time as a ranger filling them out. Our rules and policies filled four extra-deep three-ring binders. Forms and requisitions went from us to supervising rangers with lieutenant's bars on their collars, and from them to chief rangers with captain's bars, and from them to super-intendents with gold oak leaves. Our class-A jackets were festooned with gunmetal buttons and our leathers were supposed to be polished, but like cavalrymen in the hinterlands, they often got dusty. We set our digital watches to military time.

The guns that rangers carry are often thought of by the public as a recent addition, but the need for such martial protection in the face of hair-raising encounters with miscreants goes all the way back to the beginning. In 1916, the first director of the underfunded National Park Service dug into his own pocket to buy each of his rangers a pistol. Today, when the talking, cajoling, and educating are over — and all good rangers prefer these methods to the use of force — the fundamental mode of park protection remains coercive, by force of law and arms. The threats facing today's rangers are more than theoretical. According to a 2001 federal study, rangers are thirteen times more likely to be killed or injured on the job than agents of the Drug Enforcement Administration.

After the failure of the first undermanned civilian authorities, their replacement by the army, and the army's replacement by an armed and uniformed civilian police force, the problem of who would manage the parks, and how, and under what philosophy, has never gone away. Adding to the philosophical stresses within them, by the mid-twentieth century park agencies were placed in charge of an increasing number of "recreation areas" — lands of a profoundly different character from those their founders had in mind. Typical of these areas were crowded coastal beaches and the shorelines of water storage reservoirs. As any ranger who has worked in them can tell you, the atmosphere in these places is less contemplative and more boisterous than that of a nature preserve. At times it is downright lethal.

For decades park professionals have worried that the sort of duties rangers grow used to in a recreation area — controlling crowds, quelling drunken fights, and contending with an urban criminal element — would change the fundamental nature of the ranger's role. What has been less widely discussed are the effects of whole careers spent in manmade recreation area landscapes — lifeguard towers, concrete-block restrooms, parking lots, snack bars, and the muddy bathtub rings from the changing water levels of reservoirs — on the wilderness aesthetic of people in the ranger services: that love of unspoiled nature that once characterized the men and women who gravitated to park work.

Auburn State Recreation Area was one of two main areas under the administration of State Parks' American River District. The other was Folsom Lake, where the district's offices were located. By the time I came to work in the district we had a new superintendent there. Tall, blond, and athletic — he ran during his lunch hour —

Bruce Kranz was a born-again Christian with a growing family. He had started his career as a lifeguard on State Parks' Southern California beaches and, rising quickly through the ranks, worked at two other reservoirs before coming to Folsom. Folsom was basically the same sort of operation as the beaches where Kranz had started out: intensive aquatic recreation close to a major population center — in this case, Sacramento and its suburbs. Kranz was by most accounts a capable administrator of such places but had little ability as a naturalist. Nor had his assignments ever required that of him.

As it happened, Kranz arrived at Folsom right after the floods of 1986 and just in time for a long drought that followed. For the next six years the reservoir outside his office window was perennially drawn down to fill the Bureau's water contracts. It was so bad during those years that Kranz employed a full-time maintenance worker whose job was to go around the lake on a barge, setting out buoys to mark all the rocks emerging from the water so that speedboats wouldn't hit them.

By the summer of 1992 the lake's marina looked like a desert. Every day Kranz had to look at the expanding shoreline, now over half a mile of bare yellow dirt. Dust devils whirled across it, picking up beer cups and bits of paper. At night his rangers pursued kids in four-wheel-drives over it and their headlights flashed in crazy circles over the dusty wastes, as if searching for anything that lived. To make matters worse, the storm of 1986 caused a rewrite of the flood control rules for Folsom and now, at times in the winter too, almost two thirds of the lake's capacity was held empty for flood control space.

During this time Kranz developed an interest in politics. In 1992 he unsuccessfully campaigned for a seat on the board of directors of the Placer County Water Agency. Later he served as chairman of the county's Republican Central Committee. In the latter role he rubbed shoulders with Placer County's archconservative state and federal legislators, among them freshman Congressman John Doolittle, Republican from California's Fourth District, and State

Senator Tim Leslie. In such company Kranz soon become an out-spoken advocate of the Auburn Dam. Auburn, he pointed out, would store runoff from the mountains that could be used to keep Folsom full for swimming and waterskiing all summer. Further-more, Auburn would enable the flood control capacity presently held open at Folsom to be moved upstream, so Folsom could be al-lowed to fill in the winter and spring. But Kranz's motives were not entirely myopic. Like many natives of Southern California, a near-desert that by the end of the nineteenth century was reaching four hundred miles north for its water, Kranz believed that if the state was to continue to grow and prosper, we would need more water. Lots more.

So it was that in 1991, when a discussion was held at a regional meeting of district superintendents on what California State Parks's official position on the Auburn Dam should be, Kranz alone spoke in favor of drowning his own park, Auburn State Recreation Area. Within two years drought gave way to normal rainfall again, but Kranz, who once described himself as "a pro-business guy in a preservationist agency," never changed his opinion. Interviewed in 2003, he still thought a dam on the North Fork was a good idea.

If having our own boss come out in favor of putting us underwa-ter wasn't enough, new legislation to authorize completion of the Auburn Dam continued to appear in Congress. No sooner had Representative Norm Shumway's Auburn Dam Revival Act of 1987 died in a legislature more worried about deficit spending than about floods or federally subsidized water for California agribusi-ness than another Auburn Dam bill appeared in 1988. This one was a $600 million plan for a flood-control-only dam. It perished with-out ever leaving committee. But its backers didn't give up easily, and the next year they were back. Again they were defeated.

The new concept — a "dry dam" that didn't store water or gener-ate power but remained empty until a major storm, when its gates would rumble shut, filling the American River canyons with runoff for a few days or weeks — brought arguments between two fac-tions: those who supported a more expensive all-purpose dam and those asserting that flood control must be secured for Sacramento without triggering the opposition all-purpose dams had among budget conservatives and environmentalists. By 1989 local governments around Sacramento had formed their own flood control agency, and by 1992 this agency, the State Reclamation Board, and the Army Corps of Engineers were all backing the flood-control-only dam.

That year the dam loomed in Congress again in the form of a $638 million flood control project buried in a semiannual omnibus water projects bill. Again environmental groups mobilized, and the bill was defeated in a floor vote in the House by a margin of almost two to one. The following year a federal study found sections of the river in the Auburn Reservoir site eligible for designation as "wild and scenic," which would have protected it. However, no such designation was ever made. Then in 1996 the Auburn Dam was back in Congress in another omnibus water projects bill. Again environmental groups mobilized and again the dam was defeated.

In 1992, the year Folsom Lake hit bottom, John Doolittle had been elected to represent Norm Shumway's old district in Congress, replacing Shumway as the big, multipurpose Auburn Dam's principal booster. Doolittle's Fourth Congressional District was a huge chunk of sparsely populated, mountainous, northeastern California. It contained no ground at risk from the American River's floods; however, its southwestern corner happened to include three of the fastest-growing towns in California: Roseville, Rocklin, and Lincoln. By 2002 Lincoln's population increased by 28 percent in a single year, making it the state's most rapidly growing city.

John Doolittle's reelection campaigns were driven by large drafts of development money, and the water it took to sustain the kind of growth developers wanted was, if adequate for now, not limitless. So by 1995 Doolittle was vowing to use a key committee assignment and the new Republican majority in Congress to kill any solution for flood control on the American River that didn't store water and make power for his suburbs. Then January 1997 saw a storm come off the Pacific whose peak rainfalls on some portions of the northern Sierra topped even those of 1986. In the storm's aftermath, with Sacramento's congressional delegation pushing for flood control and Doolittle holding out for a multipurpose dam, the American River was rarely absent from the news. Each day we went to work in its canyons with a curse of futility hanging over us. And futility is the most debilitating thing there is for someone in a dangerous job.

A couple of months after the melee at Upper Lake Clementine I found Finch in the old mess hall where we dressed each morning, stripped down to his undershirt and green uniform jeans. On the floor in front of him was a little box about a foot high, made out of wood scraps from the maintenance shop. Finch was stepping up onto it and then back down again to the bouncy beat of a pop ditty from a couple of years before playing on a portable stereo:

> Here's a little song I wrote.
> You might want to sing it note for note.
> Don't worry, be happy.
> In every life we have some trouble,
> But when you worry you make it double.
> Don't worry, be happy.

"What are you doing?" I asked him.

"Training for our new physical fitness pay," he panted. "Sixty bucks a month if you can do this for a few minutes and keep your heart rate low enough when they take your pulse afterward."

"Seems worth it. Nice music."

"Yup. Helps me keep the pace. Sixty dollars a month — it helps."

"Sure." I nodded in agreement.

> — ain't got no cash, ain't got no style.
> Ain't got no gal to make you smile.
> Don't worry, be happy.
> Cause when you worry your face will frown,
> and that will bring everybody down.
> So don't worry, be happy.
> Don't worry, be happy now.

The song ended. Finch picked up a towel draped over the file cabinet next to the stereo. Wiping his face, he put two fingers of one hand on his neck while looking at his wristwatch. After a minute he removed them, shook his head, rewound the tape, pushed the play button, and went back to stepping up and down.

For over a quarter century — longer than anyone else — Finch would work under the waterline of the Auburn Dam. His secret, he later told me, was to keep himself busy. He went from union organizing to badge collecting, to collecting historic photos, to researching the history of rangers, to exercising with that therapeutic little song, all the while clinging to the fundamental virtue of the original idea: a ranger guarding his park and its visitors as well as he could, no matter what the politicians above him said or did.

AUTHOR'S NOTE: Finch's research led him to the idea of celebrating the 125th anniversary of Galen Clark's appointment as guardian of Yosemite. He promoted this notion until our agency adopted it, followed by the California Legislature. In 1991 yearlong observances of the 125th anniversary of the California rangers culminated in a conference at the state's oldest remaining nature park, Big Basin Redwoods — Yosemite having long since become a federal park.

Drawing on his interest in badge collecting, Finch designed a special commemorative badge for the occasion and got it approved by the department. He and O'Leary distributed the badges to rangers statewide, packaging and mailing them from our old mess hall.

In 1995, Finch self-published a history of California's state park rangers in a lavishly illustrated coffee-table book. We all bought one. Along with the anniversary celebration, the book helped us see ourselves as part of something larger and lasting. Still, none of us who worked with Finch ever let on how proud of him we were.

It wasn't our way.

9 / CROSSING THE MEKONG

For a long time the best highway maps available in California have been those published by the California Automobile Association. They are finely drawn things that presume a greater level of interest on the part of the highway traveler in wandering off the main roads than other maps do. They are known for their accuracy in mountain recreation areas. Where other maps often show generalized green blobs for parks and national forests, the auto club's depict with precision each campground, ranger outpost, secondary road, and body of water.

In the late eighties when I first worked at Auburn State Recreation Area, the Automobile Association's map of the Sacramento Valley and Sierra foothills showed a large lake filling our canyons of the North and Middle Forks of the American River, although of course no such lake existed. Theirs was not the only map that did. A Rand McNally map distributed through filling stations also showed a twenty-five-mile-long Y-shaped blue feature labeled "Auburn Reservoir" in the middle of our canyons. And the gold-mining equipment shop in Auburn still sold Metsker's Placer County map in a slipcover proclaiming THE SPORTSMEN'S GUIDE . . . NEW UP TO DATE . . . PEOPLE WOULD BE LOST WITHOUT US. It, too, showed our canyons full of water. Not surprisingly, people often came to our ranger station for directions to the lake. "You're close," I would tell them. "It's about fifty feet deep where we're standing."

I still have those maps, along with two or three hundred others. Maybe Finch with his old ranger badges, uniform insignia, and photographs of early rangers reinforced my interest in collecting. But I already had a pretty good hoard, even before I met him. I have the maps I used to find my way around in almost every place I ever worked as a ranger — mountains I climbed, rivers I ran, reaches of tundra and salt-drenched beaches and sea inlets I walked. And others, too, of all the places I traveled when I wasn't working.

My favorite are the topographic maps, the ones with all those sharply drafted lines connecting all the points on a given elevation, which to the trained eye describe mountains, canyons, ridges, and gullies so well you feel as if you could reach out and feel the high and low spots on the paper. I buy those, and all kinds of other maps, wherever I go. I even purchase, speculatively, maps of places I have no time in my schedule to explore while I am there. I keep them all in my office, on two shelves full of those cardboard document files you see in libraries. To me, they are a sort of wealth, of memory and possibility.

I have maps of the Italian Alpi Apuanni, where my second wife and I drank red wine in a little mountain hut with my friend Marco and his Italian rock climber friends and then stumbled by climber's headlamp down several kilometers of steep trail littered with slippery, just-ripened chestnuts that had been falling from the trees, and somehow made it back to his house in a village north of Viareggio late that night. And I have maps of the Juneau Icefield in Alaska, on which I wandered alone with my ice ax and crampons making sharp scraping noises in the vastness for a few days as I tried to make up my mind to divorce my first wife. In a file marked "Sierra" I have maps my immigrant father, his eyes enthusiastically open to the grandness of his new land, used in the early 1960s to navigate us around the high country wearing shoddy work boots and basketball shoes and leading a string of pack animals loaded with army surplus ponchos for tents, water bags from military life rafts, and sometimes my mother's guitar. When I come upon those, I miss my mother terribly.

One day a copy of the long-awaited new topographic trail map of the North and Middle Forks of the American River arrived in my mailbox at the ranger station, crisp and still smelling of printer's ink. I carefully unfolded it and laid it out on the table, admiring the sharpness of the brown topographic lines, the crispness of the blue line representing the river, and all the exact twists and turns of the crosshatched black line representing our boundary. There was no lake on it. And the new map somehow made this place — our strange, sacrificial park, living on borrowed time — less ephemeral, more *real*.

For years we'd used the U.S. Geological Survey 7.5-minute topographic maps. But it took a great mosaic of them — they wallpapered a whole interior wall of our ranger station — to cover our 48 miles of river in two counties. And although we had drawn our serpentine boundary in pen on the assembly on the wall, when we carried the individual maps in the field they didn't show it, and we rangers needed to know whether we were inside or outside the boundary when we heard chainsaws or rifle shots.

That wasn't the only problem with the Geologic Survey maps. Of course, the shape of the land and the bends in the river hadn't changed since the 1950s when these maps were made. But the ways people got around on the land had changed considerably. After the Bureau had taken possession of our canyons and burned down all the old ranch and mine buildings, many of the dirt roads serving them had been abandoned, washed out, or grown over. As vegetation encroached from both sides, some became footpaths. And a slew of new trails had been built, none of which were on those maps.

Then again, when I compared the new map with the old maps, there were things on the old ones I thought I'd miss. In the 1950s, the governments surveyors collected the names of places on them from local people and still older maps. At that time there were people living around these canyons who had been born only a decade after the Civil War, and only three decades after the Gold Rush had swept away the Nisenan Indian place names, along with most of the Nisenan themselves.

Many of the Gold Rush place names reflected the incredible diversity of the immigrants who came to these canyons in 1849 and, affiliated by ethnicity, religion, or origin, banded together in groups to mine gold. There were African Bar, Mormon Ravine, Kanaka Bar, China Bar, Dutch Flat, French Hill, and Spanish Dry Diggings; Maine Bar, Oregon Bar, Iowa Hill, and Illinois Canyon; New York Bar, Philadelphia Bar, and Hoboken Canyon.

The rolling divides between the canyons bore the names of families that had settled them after the Gold Rush to raise cattle: Holt Homestead, Baker Ranch, and Butcher Ranch, where the Bureau burned the last house in my time. Perhaps most evocative were the place names that remembered misfortune: Murderer's Bar, Slaughter Ravine, Small Hope Mine, Robber's Roost, Sore Finger Point, and Deadhorse Slide. Others offered hope of redemption: Honor Camp, Temperance Ravine, and Salvation Ravine.

All of these old names were missing on the new map, and in their place were some new ones. For example, the name of an Auburn banker, lumber company owner, and avid horseman who had died in 1984 was now strung out along a dotted line representing a main bridle path through the canyons. And where before the turns of the river were identified with the origins of miners, now they bore the names given to the major rapids by whitewater rafters and kayakers since the 1970s: Chunder, Parallel Parking, Tongue and Groove, Chamberlain Falls, Zig Zag, and Bogus Thunder.

At first it made me a little sad to think the traditional names would be forgotten. But then I thought, if the river is to survive, it must mean something to each generation. Those Gold Rush immigrants to which old California families trace their lineage were only the second wave — the Indians had come here perhaps twelve thousand years before — and immigration continues. Perhaps it is salutary to allow for people to bestow new meaning on these canyons periodically. We rangers had done it: the Bowl, Campsite Number One, Nude Rock, and Pig Farm were ours. Some of us weren't born here — MacGaff and Jeffries were from the East Coast. I was a native but my parents had come from Europe. And so, in this way, new Californians could make this river their own and hold it close to their hearts with the memories of what had happened along it.

One autumn day shortly after the new map came, I arranged to meet some people about a case I'd investigated the previous summer while on boat patrol at Lake Clementine. They were Laotians, new Californians. They pulled up in front of the ranger station in a pale gold sport-utility vehicle so new it still had paper plates. In California your license plates generally come within six weeks, so I knew it had been bought since July. *Did he buy it to cheer her up?*

From the first moment I saw the man and the woman, I understood. They were the kind of married people who had known each other since they were teenagers. They were in their late forties now, and would be together until they died. He sat in front next to the driver, their American son-in-law — the one who had taken the directions from me on the telephone, because his mother- and father-in-laws' English wasn't all that good. She sat in the back seat.

The three of them got out. I greeted the son-in-law and asked him how he was doing. Well, he said, it had been a tough time. I ushered them through the screen door into the ranger station. I led them to the topographic map mosaic of our canyons on one wall and showed them where we would be going, up on the North Fork. I offered her the use of the ladies' room. While we were waiting for her, I talked to her husband.

He was a slight, handsome, polite man with graying hair cut neatly and combed straight forward toward his face. He wore loose-fitting dress slacks, a button-front shirt with a subdued pattern, and nice shoes. He had worked hard and, I judged, done okay for his family.

"What do you do?" I asked.

He smiled shyly. "Mechanic," he said.

"What kind of cars?" I asked him.

"All kinds," he replied.

"Where do you work?"

"American shop."

His wife, returning from the restroom, added, "He was airplane mechanic."

"Where?" I asked.

"In my country," he replied quietly.

"What kinds of planes?"

"T-28, C-130," he said.

"C-130 — isn't that a military plane?"

"Well . . . yes ," he said, still smiling.

"Was this during the war?" I asked him.

"Yes, from 1968."

"So when the U.S. pulled out, you had to leave?"

"Yes — after a while Communists came looking for us, for helping Americans. Yes."

"Did you two know each other at that time?" I asked.

She, standing next to him, answered, "Yes, Early was born three months before we left." She smiled faintly at the affectionate joke and then explained it to me. Her husband had named their son Early in the new language he'd picked up from the Americans at the airbase, for being born several weeks after he was expected. When they had to flee she carried Early in her arms. For days they hid in the jungle with the Communists all around them. They found a man with a boat to take them across the Mekong River under cover of darkness. Eventually they came to America. Their kids had grown up solid here.

"I hear from everyone he was a very good boy," I told them.

"Yes — he want to be soldier," she replied. "He always want to help people, since he was little boy. Always help people. We lucky to have him. But he is not lucky —" Her eyes brimmed over, and she turned away.

"Yes, I'm so sorry," I said, and stood there awkwardly for a minute. The husband came over and gently touched her back, and I walked to the screen door, held it open for them, and led the way to where their car waited next to my Jeep in the shade of the tall pines.

Driving ahead, I led them across the Foresthill Bridge and up the divide between the North and Middle Forks. At the beginning of the steep dirt road to Upper Lake Clementine, I pulled over to let them pass me, so my car, not theirs, would be enveloped in dust.

We got to the bottom and stepped out of our vehicles. Down a beach of cobbles in front of us, the North Fork entered the little reservoir of Lake Clementine to our left, no more than a riverlike finger of water itself, distinguishable from the rapid above it only by its flatness. Above the inlet, the rapid was so shallow now that it was hard to imagine how the river had looked, unusually high with late-melting snow, back at the beginning of July.

Along the water the cottonwoods were beginning to turn, and their leaves sparkled as they fluttered in a gentle breeze up the canyon. A child's voice from down the lakeshore echoed off the far canyon wall. A boy fished from the bank. A lone kayaker pulled lazily for shore.

The father stayed up on the beach by the car, the son-in-law with him. The mother walked alone across the cobbles to the river's edge. She couldn't know, could she? *Right ahead of you now; that's where his body lay after he was pulled out, during CPR.* I walked down to stand next to her. There on the beach in the bright sun, in her fitted white blouse and navy blue polyester slacks, with her flat, pale, pretty face, she looked so young that I could hardly believe her son was all grown up and going into the Army National Guard.

She had a little camera in her hand. She took some pictures of the river.

After a while she looked at me, her eyes imploring, "Where — ?"

I pointed to our left. "He was sitting on the beach with the girl, his friend, there." Then, turning and pointing upstream, I told her, "They had been floating through that rapid all afternoon. They would ride it down to here, and then they'd walk back up the river-bank and ride it down again. They were all tired. Maybe he was more tired because, as his companions put it, he was 'playing life-guard.' They said he was pulling each of them out of the fast water at the lower end before it swept them into that turn, where the river stacks up on that cliff before entering the lake. He must have sensed the danger there. That's where he later disappeared."

I looked at her, she nodded slightly, and I continued. "So they were about to go home, and Early and the girl were sitting over there to our left on the beach, across from the turn where the river collides with that wall. Early saw the young boy coming down to-ward them, struggling in fast water. Early yelled to him, 'Do you need help?' but didn't wait for a reply. He stood up and tossed his car keys to the girl. She said he didn't say anything, just left her, ran to the water's edge, dived in, and swam hard for the boy. When he got within reach Early grabbed the ten-year-old and towed him to-ward the beach. The little boy later told me that when they got close to shore, Early pushed him toward several other people who had waded in to help. Now all the attention was on the rescued boy, as the people on the beach bundled him up and took him to his fa-ther. No one saw what happened to Early, who was still in the water. A few seconds later the girl he'd been sitting with noticed he was missing and began looking for him. For a moment she saw the top of his head come up over there, just downstream of where the cur-rent hits that wall. Then he disappeared, and she said he came up once more, about over there, in the lake inlet. That was the last she saw of him until they pulled him out, right here."

She walked closer, not to the place where her son had laid dead but past it, down to the river's edge. There she stood perfectly still, staring fixedly across the water toward the place I had shown her where her son had last been seen alive.

She seemed far away. She looked hard at the water, not crying, but as if she was trying to see something, as if she could peer across the months and the change in the water level since July and see him there, alone in his last moments. I felt helpless for her. I didn't know what to do. So I left her and walked back up the beach to where her husband and son-in-law still stood. Why had the husband stayed so high on the beach? I searched his face. There was no sign of anything there but a sort of stoic kindness.

I asked him how he was doing. He began telling me about his work. He and the son-in-law had been forming the foundations for an addition to the house, he said, so between that and his job he was working seven days a week. And I thought, *Yes, that's how it is with us men, isn't it? Our work is our salvation at times like this.*

But now he was looking hard at his wife and his face changed. He looked very worried. She was still facing away from us, as if in a trance. Her toes were almost touching the water; she leaned forward. I thought, *He's afraid he will lose her. She will walk around in the land of the living, but really she will be living with the boy forever at the bottom of the river, drowning in cold grief.*

The husband took one step toward his wife and called to her, urgently, in Lao.

She didn't answer and didn't give any sign she'd heard him. He called to her again, louder, his voice commanding. Again she didn't respond. He walked swiftly down to the water's edge and led her back up, firmly but ever so tenderly, by the arm. He seemed uncomfortable now. He seemed ready to leave.

"There's no rush, really. I've got all afternoon," I told him.

But he wanted to go. They both thanked me. I followed them and the son-in-law back to their car. At the car door the woman turned to face me. "On the twenty-fifth, we have food. Priests come to our house. It is a thing we must do for Early. We give food to his spirit, in case he is hungry."

Her eyes filled with tears again, and she opened the door and got in. I saw how it was with the mother — this practical kind of sorrow in which one prepared food to feed a child beyond the edge of life. As her husband started the engine, she rolled down her window and said to me, "You bring your children, you come. We have lot of food."

I thanked her and waved as they pulled away. I heard their gearbox whine up the canyon side and saw the cloud of dust behind them spread into the brush. The boy who stood in the shallows fishing and the kayaker had gone.

I was alone on the cobbles. The wind picked up, warm, dry, and dusty; it rumbled in my ears. For several minutes something like an orange haze had been passing in billows between me and the forest at the upper extent of the beach. I walked up to see what it was. As I entered the orange cloud, I saw that it was composed of thousands of migrating lady beetles, an endless swarm of them, undulating up the canyon on the wind. They collided with my hat, making little ticking sounds. They landed on my uniform. They crawled on my sleeves. I let them wash over me, facing the direction they were coming from. The autumn afternoon sun shone through a million wings, turning the light the color of saffron, like the robes of Buddhist monks.

I heard a motor. A young man and young woman arrived at the bottom of the road by the beach in an old white car. I waved them down. I showed them the little beetles on their windshield and told them I had been standing there watching the swarm go by unabated on the wind for twenty minutes. I told them I had worked this section of the river for more than ten years and had never seen anything like it.

I looked at them. They were wearing bathing suits, sitting on towels on the hot seats. The young man sat quietly, respectfully. He had turned off the car now. They had both listened to my little natural history talk without replying. Could they go now? their expressions seemed to say. I looked down and realized I was wearing a badge and a gun. I smiled sheepishly and waved them along. The young man started the motor and left right away, and the car disappeared into the willows down the sandy track along the lakeshore. I was alone again. The saffron light and the brushing of the beetles against my face, my hands, remained.

The cold force of the American River that July has long since spent itself on the ocean. I haven't seen a swarm of migrating lady beetles that big since then; maybe I never will. My friend Marco got remarried and now lives in a different village. I don't know if the old stonecutter who poured his wine so freely in the hut Marco took us to in the mountains is still alive. In life, you travel and what is behind you keeps changing after you are gone.

But on the maps I keep, the places and the stories of people I met there don't disappear. It is also like that with the memories of things that happened to me as a ranger. Some things that happened years ago still raise my heart rate when I think of them; the difference is now I know what happened in the end. I remember the urgency as if I were inside it again. But I also see it at a great distance, like looking down on the tiny mountains on a topographic map. And in a way everything seems to be settled now, and in another way it never will be.

I am in the jet boat, going desperately fast upstream where Lake Clementine grows shallow and turns into a river. The siren is screaming. I'm yelling in my headset to the dispatcher about getting a rescue helicopter up there as the V-8 thunders underneath me. The canyon turns east and narrows, and in the shade of its depth the blue strobe overhead flashes on my deckhand's face as the boat pitches on its wake. He's braced himself into the V of the bows with one hand on the gunwale and the other on the diamond-plate bulkhead. I've sent him forward to watch for sandbars, which at this speed could kill both of us if we hit one. In the din we communicate about this matter with glances and gestures.

The helicopter lands; the gray-blue body is loaded by men in blue jumpsuits like angels. It flies away.

I am swimming underwater at the lake inlet, tracing the last moments of the deceased for my death investigation. My deckhand watches anxiously from the boat tethered to the riverbank above. Fighting the cold July current, I claw my way along a submerged face of rock, groping the bottom for a place where the strength of the water might have stolen and kept a human body.

I am on the phone with Mrs. Ditsavong. My children have the

flu; I'm sorry but I can't come to the service today. On the other end of the line, I hear the chanting of the monks and the drone of a harmonium, deep and ancient.

Mr. Ditsavong is out on the tarmac, working in the hot sun of Southeast Asia, wiping his hands on a rag. The jets whine in the background.

The saffron beetles caress my skin.

In the jungle, Mrs. Ditsavong is always holding her little boy so close. They get into the boat; they sit; the little boy is falling asleep now. The boat leaves the shore into the moonless night. The boatman leans into his sculls and she can hear the drips from the ends of the oars as they are lifted out of the water and slipped back in as quietly as the boatman can manage. And they leave the still water along the bank and go out into the great current, crossing the Mekong.

10 / AS WEAK AS WATER

EARLY ONE AUGUST MORNING Will Reich and I loaded our oar boat onto the park's Ford six-pack for a whitewater raft patrol of the upper Middle Fork. He was a new ranger, a big, barrel-chested, lantern-jawed man in his early forties with a shock of unruly black hair. As usual, our job on the river that day was to monitor the guide services offering excursions by whitewater raft in our canyons, and to be available to help in the event of a boat accident. But on this day we also planned to call on a miner who'd claimed as his own fiefdom a remote beach the guides and their clients used as a campsite.

The forty-eight miles of the American River lying beneath the waterline of the endlessly delayed Auburn Reservoir had long been withdrawn from mineral status; the Bureau of Reclamation wanted its title to the land unencumbered when it was finally allowed to flood it. That meant you could look for gold but you couldn't stake a claim to the land it was on under that antique provision of western law that still allowed a few rugged individualists to appropriate a piece of public land, scar it in their attempts to make it yield precious metals, and live on it indefinitely, rent free. Nevertheless, this man seemed to believe he had claimed the beach at the mouth of Otter Creek, and the word on the river was that he might be a taco or two short of a combination plate. He had a horse and a mule down there, and he'd evidently used them to pack a considerable quantity of gear and garbage into the canyon. Will had been sending him registered letters at a post office box in Georgetown, but they'd all gone unanswered. So today we'd serve him his walking papers in person. If he balked, we might write him a ticket. We couldn't arrest him: We'd be out of radio range for most of the sixteen-mile run, so there was no getting a helicopter, and it wouldn't do to run rapids with a man in handcuffs. Then again, he didn't know that.

We left the ranger station and drove east up the Foresthill Divide. At the town of Foresthill we turned southeast on Mosquito Ridge Road, descending the two-thousand-foot north wall of the Middle Fork toward the river. Shafts of morning sun from between the high peaks up-canyon gilded each prominence in the velvety knob-cone pine forests along the far canyon wall. We passed the portal of a hard-rock mine. Air hoses for drills ran into it, but the gate across the entrance was padlocked. Hard-rock gold mining had become a romantic anachronism and knobcone pines were considered worthless by loggers. Water and electrical power were what connected this canyon to the economy now.

By the early 1950s, the race was on throughout the West. Whatever waters still ran listlessly into the sea were to be harnessed to do work and make wealth. With construction underway at Folsom Dam on the river's main stem, it became clear to the businessmen and politicians of Placer County that under the Gold Rush–era law of "prior appropriation" — whoever gets there first gets the water — the federals were going to own all of Placer County's water and power rights if the county didn't claim them first.

So Placer County moved aggressively to grab the upper Middle Fork. In 1961 local voters underwrote a $140 million bond issue to begin building what was probably the most ambitious water and power system ever devised by a nonmetropolitan county in California. When a huge storm in 1964 caused the failure of the county's partially complete Hell Hole Dam, destroying five bridges, thousands of trees, and a hundred-foot-long bucket-line gold dredge downstream, it didn't stop the Water Agency. Completed in 1967, the Middle Fork Project gathered rain and snowmelt from a 429-square-mile watershed into 7 reservoirs with an aggregate capacity of 342,000 acre-feet. From the reservoirs the water traveled through 24 miles of tunnel and over 3,600 feet of penstocks — big pipes plunging down steep canyon walls to develop the hydraulic pressure it takes to make electricity — into 4 hydroelectric power stations on the way down the mountains. Today we'd be launching our boat at Oxbow Powerplant where the river was allowed to become a river again until it reached Folsom Lake.

The last pitch of road switchbacked down cliffs of dark slate to a dirt parking lot on the downstream side of the dam supplying Oxbow Powerplant, called Ralston Afterbay. There, Will and I unloaded our boat, its aluminum rowing frame, three oars (one for a spare), two paddles, dry bags full of gear, food, and a medical kit, cam straps, helmets, rescue vests, and ammo boxes and then carried them down a dirt track to the water. The track ended at a little pool at the base of a three-story concrete wall between mossy cliffs of dark stone. A thousand cubic feet of water every second surged from a hole in the base of the wall. Drawn from the cold bottom of the reservoir upstream, it had passed through a gleaming underworld of electrical generators hidden in the canyon wall behind us, where it had just been subjected to unspeakable violence in the turbines' scroll cases. It came to light again in this small pool, as cold as ice and as clear as gin. In the three decades since the powerplant was completed, willows had established themselves around the little pool, so that it seemed to be some combination of a manmade and a wild place.

One day when I had been preparing to get on the river, I had happened to encounter the maintenance supervisor for the powerplant as he came up from underground. I spent a few minutes asking him how the powerplant worked. During this conversation, he told me that his men were working on an eroded turbine blade.

Water and the tiny bits of sand it carried, he said, can literally wear away steel. So the hardened steel blades of the turbines had to be periodically restored to their original contours by welding new metal onto their leading edges and grinding it to shape.

On another occasion, an engineer for the Bureau of Reclamation at Folsom Dam had shown me photographs of some of his colleagues standing in one of the huge tunnels through which water travels to and from the turbines inside a dam. The hard-hatted men were examining enormous galls in the tunnel's walls and floor. The engineer explained that rushing water can quickly erode even the high-grade steel-reinforced concrete of which dams are made. Unless the tunnels are constantly inspected and air bubbles are properly mixed into the water to provide a cushion, the water can literally eat a dam's interior, he said. The Chinese sage Lao Tzu was born 2,400 years before anyone thought of capturing rivers behind towering dams, but he nevertheless seems to have understood what would happen to them without constant attention:

> Nothing in the world is as soft, as weak, as water; nothing else can wear away the hard, the strong, and remain unaltered.

As we reached the powerplant's outflow one guided party was just paddling away on it: four sky-blue rafts full of matching yellow helmets, orange lifejackets, blue wetsuits, and nervously excited whoops. In recognition of the income the guide industry brought to the motels, restaurants, and private campgrounds of Placer and El Dorado Counties, the powerplant's operators had agreed to provide reliable flows in the river on weekends. Their releases usually tapered off by midafternoon, when boaters were well downstream. The commercial rafters liked to get on the river early, so they could give their clients time to lounge on hot rocks during lunch and still stay on the crest of the release as it ran down the canyon. We could move more quickly, so we generally started later and often passed them at lunchtime.

In the quiet after their departure we slid our boat into the water.

It came alive, dancing and tugging at the bowline in my hand. I tied the line to a willow branch and jumped in. The weight of my feet on the inflated floor caused cold water to well up through the lacing at its perimeter and run over the tops of my wetsuit booties, chilling my feet. The water also cooled the air in the raft's inflation chambers, and I felt the boat go flaccid around me. Will came aboard with a barrel pump, and we connected it to the valves on the various chambers, topping off each one until the boat was firm again. I went ashore and began passing him our dry bags, and he lashed them down with a handful of bright nylon cam straps.

If a whitewater boat is the domain of wilderness romantics without real jobs, it is an awfully serious, businesslike one. Everything must be tied down securely so it won't be lost if the boat is swamped, capsized, folded in half, or flung through the air by the powerful hydraulics of the rapids. No ropes or straps shall be left in a tangle where they might snag a leg, arm, or neck during a capsizing. Will and I attended to our work as if it mattered. We checked our guns, made of stainless steel so they wouldn't rust, then secured the gun belts in two waterproof steel boxes on either side of the oarsman's seat. We pulled our life vests and the chin straps of our helmets tight. I clicked the release on the diver's knife attached to my flotation vest to make sure I could get it out. Should the boat flip over and you become tangled in lines or pinned between the boat and an underwater rock, the knife was for cutting your way out — right through the boat, if necessary. Finally I checked the carabiners on the throw bags full of line for rescuing a swimmer if someone went overboard.

The first part of the river was not, strictly speaking, a river at all, but a groove blasted into ledges of slate from the outlet of the power plant back to the river's original channel, now dry below the dam. Yet its banks had grown over with alders and were now littered with river rocks and drift sand from high water, so it looked like a wild place. The boat collided with a series of standing waves, cutting off the wave crests into our laps. The water was cold enough that it caused us to suck in our breath. That involuntary response could get you in trouble if you fell in.

We entered the river's original channel, passed some abandoned placer mines, and came to a narrow jeep road bulldozed down to the river through the alders. This road had been the center of a property-line dispute between the man with the bulldozer and the man who claimed to own the land. The argument had been settled with guns, or so went the word on the river.

Our first major obstacle was a rapid called Tunnel Chute, a feature as manufactured as our put-in. Here, over eons, the river had cut a sharp oxbow until it passed close by itself a mile downstream, with only a tall fin of rock between the two channels. As the Bureau would later do at the dam site, nineteenth-century miners figured that if they could shortcut the river through that fin, they could dry out the oxbow and plunder its bed for gold. They accomplished this job with blasting powder, mules, and hand tools, and today the river runs in a tunnel through a couple of hundred feet of rock. But the shortcut made the river's pitch steeper, and no matter what you try to do to them, rivers adjust themselves according to their own laws. In response to the short-circuiting of its slope, the Middle Fork began scouring its bed on the upstream side of the tunnel until it excavated a ledge of bedrock. So blocked, the river went back to its old ways in the oxbow. In response, the miners blasted a channel through the bedrock ledge to drop the river back into the tunnel. This channel was about eighteen feet wide and fell away at a steepness roughly equivalent to that of the front steps of a public building. Through it thundered the river's entire one thousand cubic feet of water each second. The miners had never intended anyone to go through it in a boat.

To do this, you pulled over to the riverbank, lashed your oars lengthwise to your rowing frame, tightened your gear lashings, and snugged your helmet and lifejacket again, just for good measure. Then, using your canoe paddles, you took the boat out into midstream and over a drop just upstream of the main event. You then crossed a quiet pool, which gave you enough time to wonder why you were there as you positioned yourself for entry into the Tunnel Chute. Once you dropped into the chute itself, you sat down in the floor of your boat, wedged yourself in as well as you could, and hung on for dear life. Somewhere between a few seconds and an eternity later, what would happen had happened. It was so violent a ride it was really beyond anyone's control. However, most everything and everyone washed through it; it didn't trap people, just occasionally broke their arms and legs. Once I was knocked out of my boat at the top of it and had to swim it. The sheer speed and limb-tearing power of the froth — not water, but a blend of water and air too light to float in yet too wet to breathe — were horrifying. I remembered to roll into a tight ball. Collisions with a couple of boulders left bruises on my thighs and back that lasted for weeks. When it was over, I surfaced facing upstream in the echoing green tunnel with bubbles from the waterfall rising languorously around me. The first thing I saw was the white wall I had just come down in the brightness outside the mouth of the tunnel. For a couple of minutes I couldn't seem to get enough air. I was like a starving man at a banquet.

Beyond the Tunnel Chute we ran the Three Queens, Kanaka Falls, and Cache Rock Rapids, then several smaller, unnamed ones. Better that they remain unnamed, I thought. Even the well-known rapids for which whitewater guidebooks give step-by-step instructions are periodically rearranged by floodwaters, and then they become mysterious again for a few wonderful months of the following spring. It isn't good for the world to become too well known, we whitewater rangers think. Said Lao Tzu:

> The way you can go
> isn't the real way.
> The name you can say
> Isn't the real name.
> Heaven and earth
> Begin in the unnamed.

By now we'd allowed the commercial rafts to gain a good mile or two on us, and they'd long since passed from view around the Middle Fork's stately bends. As we floated deeper into the canyon, the effect was the opposite of climbing a mountain. On a mountaintop the whole world is laid out beneath you, but in the bottom of the Middle Fork the whole world is gone; there's just moving water, green and brown canyon walls, and sky. The river remained unmarked by those ahead of us and for all you could see we could have been the first ever to run it, or the last two men on earth. The late Tang dynasty poet Chia Tao was born twelve hundred years before the invention of Hypalon, the strong, flexible fabric out of which our craft was made, yet he understood how a river is always new to each one who runs it:

> Passing on the river, a boat leaves no trace on the waves.

By the first week of August it was hot in the Central Valley. All visible snow was gone from the high peaks up-canyon and white towers of thunderhead rose above them most afternoons. By the time the Middle Fork's waters bore us through the gorge, they had already kept air conditioners humming somewhere in the great civilization down-canyon, to the west. Before that those waters had been snow on the mountains, and before that clouds, and before that part of an ocean around Hawaii or the Gulf of Alaska. The winds that would bring them back around the cycle, or not, and the snowfields that would shrink and disappear over the next hundred years, or not, would now do so or not according to the aggregate actions of bankers and oilmen in Houston, Almaty, and Bahrain, and of politicians and motorists in Washington, D.C., Berlin, Paris, and Beijing.

From this bewildering world to these fabled canyons of the Gold Rush continued to come a small stream of those whose fantasy was to go into the mountains and live like nineteenth-century miners. In a way they were no different from us rangers, for we too had sought our own version of the simple life here on the edges of civilization — if not, it turned out, beyond its reach — or from the whitewater guides who sought their own simple peace in the exigency of the right command at the right moment. "Paddle hard, forward, now!" we heard them yell at their clients, who had paid good money to be yelled at — their own version of simplicity — as they entered the bigger rapids.

The river turned south. The Hornblende Mountains loomed on our left, steep green battlements of pine and Douglas fir. The river was quiet in this section, a sinuous path of silver reflecting the afternoon sun between bright green ranks of willow and alder at the water's edge. Helmets off, wetsuits off, wearing only shorts, we spelled each other on the oars, dabbed our noses and ears with sunscreen, and ran an easy series of riffles between long blue-green pools. We passed African Bar in contented silence, pulling our shifts at the oars.

Around another bend, Otter Creek entered the Middle Fork from our left at such a sharp downstream angle that the ridge between it and the main canyon had eroded to a cleaver. The ridge top was so sharp it appeared not wide enough to walk on, its sides near-vertical jungles of hardwoods. It rose steeply to an old jeep road on Cock Robin Point, sixteen hundred feet above us, and it was from there that our man had apparently made his way down with his horse and mule. You had to admire his fearlessness, or pity his foolishness, as the case may be. His camp lay on a delta of sand that Otter Creek had deposited around its junction with the Middle Fork. Nothing moved there as we approached, and there was no smoke. We rowed in and pulled our boat up on the beach. In front of us, nailed to a post made of a rough branch, was a hand-lettered sign on a piece of cardboard: THIS IS A MINING CLAIM. DON'T THROW YOUR GARBAGE ALL OVER OR ELSE.

We crossed the beach to the edge of the forest where a tent stood in the shade of an overhanging bay laurel. A rock fire ring had been constructed next to it. The ashes were cold to the touch. Around it were a couple of cheap pots and a few dishes, a gold pan, an aluminum sluice box, and a five-gallon bucket partially full of black sand

and water — the last stage of the refining process by which placer miners recover gold dust.

Hoof prints could be seen everywhere, but there was no sign of hay and not a kernel of grain. There wasn't much for a horse or mule to eat in the surrounding woods but thorny buckbrush, wilted redbud, and the less spiny leaves on the tips of scrub oaks. The animals had trampled and torn at every inch of the adjoining canyon wall in their search for nutrition.

Inspecting the tent, I soon found a rent in one side big enough to walk through at a crouch. Peering through the shredded fabric I found the interior in a shambles: a disheveled sleeping bag, clothes, scraps of paper, and various personal items were scattered about, and over them had been spread a white dust like flour, the contents of several bags of macaroni, and some kind of dried goo — tomato sauce or beans. It looked like there'd been a fight.

Will and I split up to search for survivors. I went up Otter Creek, Will downstream along the river.

Otter Creek's canyon was as deep as the main one but far narrower. Its walls were even more precipitous than the Middle Fork's and its bottom so close and junglelike that the only practical way up it was to wade in the creek. The moment I left the beach and entered the shallow water, the world was reduced to a colonnade of gray alder trunks, their canopies obliterating even the narrow strip of sky visible between the canyon walls. The aloneness was intense. The creek sparkled like quicksilver in the dim light, and on either bank great tropical leaves of Indian rhubarb grew from corms that clung to the mossy rocks like the gnarled fingers of an old man's hand. There was no other world. After a quarter mile of wading I had found no trace of the miner or his animals, so I started back.

Back at the camp there was no sign of Will. I climbed through the hole in the tent to search for clues to what had happened among the disordered contents. I came upon a notebook, densely inscribed in a small hand. It was like a shopping list, or perhaps a journal from the manic phase of a bipolar life:

> Get topographic maps . . . set up solar battery charger . . . learn tan-
> ning for hides . . . latigo for saddle . . . ammo . . . needles and thread
> . . . buttons . . . get some books on minerals to read . . . get fish
> hooks . . . what kind of fish?

I leafed though it. More of the same. Near the back I came across a
single observation, no doubt reflecting the miner's impression of
the way into the canyon:

> Trails not for the faint of heart.

Will came back up the beach, having found nothing. Leaving
him to have one more look around the camp, I struck out across
the beach to the mouth of Otter Creek. Halfway there I came upon
an explanation for the torn tent, the mess inside, and the miner's
apparent disappearance: a huge pile of fresh bear scat of a surprisingly large diameter. It seemed our intervention was no longer required here: A bear had served our eviction notice. Perhaps it had
occurred at night, for bears tend to be nocturnal in the hot summer
months. Thankfully there was no blood, but the encounter had no
doubt left an impression on our man, for he'd obviously departed
in a hurry, leaving his entire camp to the bear.

We got back into the boat. The remainder of that day's patrol
passed without incident. Weeks went by, and the camp remained
abandoned. In September Will and a couple of his seasonal aides
went in and cleaned the place up, burned what they could, and carried the rest out in their raft. The miner was never heard of again
on the American River.

Our other river, the North Fork, was a natural flow stream, its
main channel undammed above Lake Clementine. The rafting and
kayaking season began when the North Fork swelled with rain in
January and continued as the snow in the high country melted in
late spring. Then, as flows on the waters of the North Fork went
down, boaters moved to the dam-release Middle Fork for the remainder of the summer.

The commercial guide industry on the American River had begun in the late 1970s, when young men and women who had been spending a lot of time on the river decided to try to make a living taking other people down it. Few wilderness activities are better suited for a guide business than whitewater rafting, in which novices lack both the equipment and the knowledge to negotiate a river safely yet, when properly outfitted and guided, are usually wildly enthusiastic about the experience. In those days any competent boater who could put together a raft, a few paddles and helmets, and an old school bus to get his or her clients to the river could clear $400 a weekend — not bad for people who'd been living in vans, following the spring melt-off around the mountains of the West.

The American's South Fork, an easier run that soon became the most heavily used whitewater river in California, lured people to try the more challenging North and Middle Forks. In 1982 rangers issued the first six permits for guide services to operate in Auburn State Recreation Area. Three years later, fifty-seven companies were offering whitewater trips on our rivers. A 1985 study showed that paddlers spent nineteen thousand person-days annually beneath the waterline of the Auburn Reservoir. Now rangers were called on to handle whitewater raft accidents and rescues. Hostilities flared when the rafters got out of their boats in places the miners saw as their own, or when incensed boaters cut ropes stretched across the river at neck level, which the miners used to tether their dredges. By 1986 Ranger Sherm Jeffries was writing a management plan for whitewater boating on the American, and over the following couple of years some of us rangers enrolled in guide schools and whitewater rescue courses. Eventually we got our own raft.

Will Reich had come to work for us in the nineties as a seasonal boatman. He'd previously worked as a guide and was better at running a boat in swift water than any of the rest of us. Eventually he went to our academy and came back as our river patrol ranger.

On another patrol that summer, Will and I stopped for lunch downstream at Dardanelles Creek. There we munched our sandwiches reflectively, seated on steep, warm sheets of polished bedrock plunging into a deep pool where our boat bobbed, tied off at the riverbank beneath us. When we finished eating I checked the bowline and splashed some cold water on the chambers so the air expanding in the hot sun wouldn't pop our raft. Then Will and I climbed up the canyon wall, following Dardanelles Creek into a slot canyon so narrow you could touch both sides at the same time. The dark stone was water-polished into sensuous curves. The air was cool. We waded through a series of potholes to the base of a small waterfall.

Two hundred and fifty feet beneath the waterline of the Auburn Dam, we stopped and looked up. Tufts of fern growing from the walls and overhanging branches of trees on either side of the slot above were silhouetted against a thin slice of sky. A thread of water splashed down next to us, throwing cool droplets on our faces. Then, on a little ledge just above head level, we caught sight of a green orb the size and shape of a large cantaloupe, intricately woven from moss. One side featured a small round doorway. It was the nest of a water ouzel, a bird we often saw darting beneath the rapids in search of aquatic invertebrates and fingerling fish. Like the canyon wren the ouzel was a creature of moving water, not reservoir lakeshores. It was the first time I'd ever seen one of their nests, which are usually well hidden.

In the practice of wildlife management there is a theory, called mitigation, for dealing with the loss of wildlife habitat when reservoirs are built. Mitigation means you do what you can to improve habitat on the land surrounding the inundated place, to make up for the loss. The mitigation package for Auburn Dam included the periodic burning of mature brush fields to spur the growth of new, tender shoots, which make good forage for deer.

But there was nothing in the package for water ouzels.

When our lunch stop was over, Will and I got back into the boat.

That morning most of the bright-colored flotillas of commercial rafts had been ahead of us, but, rowing vigorously, we'd passed many of them before lunch.

As we set off again, the first odd thing we noticed was a growing line of wet rock above the water's surface.

"Does it look to you like the flow's dropping off?" I asked Will.

"I was thinking the same thing — but it's way too early," he replied, pulling at the oars.

The next series of rapids, known collectively as Ruck-a-Chucky, were still a ways downriver. But for several miles now you could tell the river was up to something — hoarding altitude, its rapids nothing more than fast little riffles between long pools where the water was so sluggish you had to row against an afternoon headwind or you'd float back upstream. At Ruck-a-Chucky, the unstable walls of the canyon had been calving off into the river for tens of thousands of years, forming a natural dam and a series of waterfalls through huge boulders. Ruck-a-Chucky announced its presence at the end of a long reflecting pool where Canyon Creek tumbled down the left canyon wall through dense stands of fir. At the end of this pool the canyon narrowed and bent sharply to the right. From around the bend came an ominous rumble.

I was rowing across this pool when I noticed I had to thread my way through the sandbars or run aground. On a beach to my right I saw a line of wet sand above the present waterline. Will put his helmet back on and took the oars while I strapped on mine. Just ahead of us was what boaters call a "horizon line" — a place where the visible surface of the water ends abruptly, which means there's a waterfall. Will lined the boat up just right and we dropped over the edge as usual. But at the bottom of the falls our bow struck hard against an unfamiliar rock sticking out of the water, stopping us dead for a few seconds, perched at a steep angle. Then the impatient water picked up the boat and lifted us past the obstacle.

"That was different!" I yelled to Will above the foam.

"No question about it — the flow's really decreasing now!" he yelled back. He looked worried.

Will stroked over to the right wall, where we tied up to a boulder to ready ourselves for the process of getting our boat through the main rapid. We'd never learn exactly what happened at the power plant; perhaps somewhere on the grid a circuit breaker had popped or a turbine bearing had gotten hot, or someone had just thrown the wrong switch by mistake. But for whatever reason, our river had been turned down at the power plant to just over a third of its normal flow. With the water stretched out through miles of canyon, it took a while for the effects of this to reach us, but that was now happening.

At the main Ruck-a-Chucky rapid, the water ran over, through, and under a pile of house-sized gray-green boulders. This waterfall was only occasionally run by the certifiably insane. The formula here was simple: If you swim, you die.

Over the years a variety of approaches had been employed to get boats and gear around Ruck-a-Chucky. In 1986, when I first worked on the American, people unloaded their gear and lowered the empty rafts down the rocks next to the falls with ropes. Later State Parks constructed a portage trail and boats were carried around. Finally, we all learned to "ghost boat" the falls.

Ghost boating was at minimum a two-person operation. It worked like this: I'd hike around the falls with a single paddle in my hand. Below the falls I'd wade into a deep green pool and swim across a narrow channel to the steeply inclined downstream face of a massive boulder in midstream. I'd climb that until I stood on its top, facing upstream, ten or twelve feet above the rapids. In front of me there was only one refuge from the fast water: an eddy on the downstream side of another boulder. I had to leap far enough out into the current to land perfectly in that patch of protected water while holding on to my paddle. When I surfaced I'd grab for the boulder, then climb its downstream face until I stood on top of it, six feet or so off the rapids. Once in position there, I'd blow a whistle clipped to my life vest and tap my helmet with one hand to signal Will, who watched my progress from the top of the falls upstream.

Will would then disappear, remove some of our gear from the boat to be carried around, and push the unmanned boat out into the current above the falls. Within a couple of minutes the underside of the boat's bow would appear at the brink, hesitate for an instant as if in fear, and plunge over the falls. At the bottom, the self-bailing boat — it had an inflatable floor with little scuppers around its perimeter for water to drain out — would bob from the froth and follow the main current through a boulder garden upstream of me, hidden by the rocks. Suddenly the craft would reappear where it now had to pass between my boulder and the canyon wall, a space only inches wider than the boat itself. My job was to leap into it as it passed underneath me, paddle feverishly to bring it under control, and maneuver it into a quiet inlet between the towering boulders to my right, where we'd reload the gear Will was carrying around. If I missed this cove I could look forward to running the next rapid alone with a single paddle and no oars.

But on this day we never got that far. I set up, signaled Will, and waited as usual. He disappeared, the minutes passed — and no boat. Eventually he reappeared at the top of the falls, and I heard his whistle above the roar and saw him signaling me to come back upstream. The only way back was to jump into the rapids and swim for shore. I did so.

Once I rejoined Will at the top of the falls, I saw our boat hung up sideways on a boulder at the brink. There was no reaching it.

It didn't seem dignified to abandon our vessel to be salvaged by some lucky river rat downstream when the water came back up. State and federal budgets being what they were, we might never get another. And of course we'd become famous in a way a ranger doesn't want to — like two rangers I knew whose handcuffed suspect somehow slipped his cuffs, climbed into the front seat of their patrol car, and drove away. The pursuit that ensued involved several other rangers and a number of sheriff's cars.

So we decided on a belayed swim. Was it really a good plan? Only if we survived it. I tied one end of our rescue rope to the D-ring on the back of Will's rescue vest. As a last resort, if he got pinned by the current and I was unable to pull him back, he could release his belt, but he'd be unlikely to survive the falls if he did. In order for it all to work I just had to be a really great belayer and he had to be a really great swimmer. Ranger work brings out the best in people.

Holding a loop I'd tied in the other end of the rope, I swam out to a flat-topped boulder in the pool upstream of the falls and set up my belay on top, where a prominence gave me a good foothold. When I was ready, I pulled up slack and Will let the current and the taut rope pendulum him out into midstream. Then he began to float toward the falls, as I paid out a little rope at a time. When he got close to the boat on the verge of the falls, I felt the current begin to tug him harder. Will signaled for more rope. I wasn't sure whether his wife would appreciate it if I gave it to him. But I wanted our raft back, so I did.

A yard or two more and Will caught hold of the raft. He signaled urgently. I started pulling him and the boat back upstream. To do this I stood facing him downstream and leaned backward against the rope around my waist with my legs bent. Then, straining against Will, the boat, and the current, I straightened my legs. When I had stood all the way up, I squatted quickly, catching up the slack I'd created before the river could take it back. After a few minutes of grunting and sweating, I had dragged Will far enough from the falls that he could swim the boat safely over to shore. I swam back in.

While we had been intent on retrieving our raft, the whole Ruck-a-Chucky rapids had stopped working for whitewater boating. One group had hung up in the first rapid where we'd struck a rock but washed through. Water poured through the craft's interior and its passengers clung to surrounding rocks as their guides tried frantically to rescue them. With that rapid blocked other parties couldn't get through at all. Meanwhile, I thought, somewhere deep in the control rooms of dams and electric grids, men and women were sitting in front of angled panels on which little LEDs blinked on in orderly branched schematics and blue computer monitors showed nothing was wrong. They were probably listening to Rush Limbaugh and on their desks cups of coffee were turning cool in mugs with inscriptions like "Western Power Administration Conference 1997" and "World's Greatest Dad." Could they have any idea what the flip of a switch could do to us here? Probably not. The world was not founded upon such empathy and imagination.

Although it is counterintuitive, a rapid can be far more dangerous at low flows than at higher ones. At low water, rocks you'd normally wash over stick out and try to grab you. The water runs through them like mouthwash through the gaps of your teeth, straining out boats and the boaters who fall out of them, to be pinned underwater and drowned.

We hiked up to warn the guides for the parties stuck upstream that it had become too dangerous to run Ruck-a-Chucky under these conditions. Luckily, they had managed to pull all of their clients from the river. At this bend in the canyon our radios began working again. An old jeep road ran up the canyon wall above the rapids. We radioed Will's seasonal helpers, who had picked up our truck and driven it around to wait for us, to make arrangements with the outfitters' van drivers — also waiting there — to evacuate their clients by van on this road. Then, with our seasonals stuck directing traffic and darkness coming, Will and I decided to try to run the rest of the river. We set out, lining the boat over the falls like the old days. Then we got in.

A quarter mile downstream from the main rapids we hung up in another drop. Once firmly lodged, our boat became an obstacle to the current, which then flowed over and through it. We were up to our waists in foam. We clung to the boat, trying not to get washed out. I started laughing, giggling. Will looked at me questioningly, then his sunburned face broke into a grin. I fought my way upstream through the froth to where the boat's stern protruded from the water. Once there I began to jump up and down on it like an ape. The boat groaned; a dull scraping noise resounded through it; we felt it budge and then hold fast again. I jumped some more; Will caught an oar in the current and we were off, with me still laughing and jumping like a madman and Will chuckling as he pulled on the oars.

It was evening. Downstream from the Ruck-a-Chucky rapids we floated through a series of long, deep pools. In a canyon it grows dark from the bottom up, and close in now, the water and the cliffs on either side of us were wrapped in indigo. Framed in this dark V, a portion of the canyon wall upstream was still lit brilliant orange by the last rays of the setting sun. Then the sky dimmed, the pool beneath us went inky, and the first star reflected off it. The call of a canyon wren echoed down the cliffs.

This river and its canyon were no longer totally wild, nor were they entirely manmade. Rather, like much of the rest of the world, they had become some mixture of the two. No part of the world could now be said to be entirely untouched or unaltered by human enterprise; radioisotopes had been found in Arctic lichens, ice shelves were falling off the Antarctic cap, and it now looked as if the weather itself could no longer be considered entirely "natural." Citing these facts, some university intellectuals had concluded that there really wasn't any "nature" or "wilderness" anymore. Further, considering what we were now learning about the aboriginal use of fire to manipulate ecosystems, the fact that some so-called hunter-gatherer cultures actually cultivated wild plants, and the possibility that prehistoric hunters may have had something to do with the disappearance of some Ice Age animals, maybe there never had been such purity — at least as far back as human culture existed. Therefore, said some intellectuals, any moral claim you could make for saving what was left of wild, unregulated nature was based upon a faulty premise or, worse yet, pure sentimentalism.

Under the powerful influence of postmodernism's cultural and moral relativism and an almost excretory, childlike pride in human creations — our bioengineered crops and animals, brainlike computers, and the Internet — some of these thinkers had gone so far as to say that we ought to finish the job of domesticating the earth and yoking all of it to productive purposes. One writer even claimed that the planet's physical and biological self-regulation is now being replaced by electric grids and communications networks that, with the intimate involvement of human beings, will become the earth's new nervous system.

We rangers have a fair amount of time to read and I'd been aware of these ideas for a while. They are merely a more fashionable version of traditional human-centered technological optimism. But seen from a boat on a regulated river that night, the claims of these postmodernists looked faulty. However poorly managed that day, the job of metering a single river to generate power without killing any whitewater rafters was far simpler than managing the climate that provided the river's water. If dams had many beneficial effects for civilization — our late-summer whitewater rafting season being one of them — they also had many unintentional outcomes. Coastal beaches were now deprived of their sand, for centuries replenished by rivers wearing down mountains. Some of the beaches would now grow rocky — and that change might have an effect on, say, the economy of a beach town or the nesting of plovers, and that change still another effect. In California, a wild salmon fishery so robust that even after the Gold Rush people were still pitchforking fish out of Central Valley streams had been nearly done away with by dams. We humans were reductionists, and neither our brains nor our most powerful computers can begin to account for the complex web of interrelationships in a global ecosystem.

In the end much of what is seemingly known and tamed is in fact unknown, and untamed. Even with our interventions, and now

because of them, the world continues to be mysterious and acci-
dental. There are surprises in its most compromised corners, where
water is eating dams, lichens establish themselves on concrete, willows take root, and we can almost get killed in wilderness sport because someone turned off the river. It may well turn out to be a
more dangerous world for all our efforts to domesticate it. We have
always been the beneficiaries of nature's largesse and we take all of
this for granted, as adolescents do their parents' roof over their
heads, and now we want the car keys. But I, one ranger, do not.

This evening on the river was an achingly beautiful one, and not
because of anything I or my species had done; I did not make the
uplifting of mountains, the endless wearing-down of them by water, or the adagio of the canyon wren, nor do I want to. I, one
ranger, want only for the unregulated wild that has always provided
for us to outlive me and all my progeny. While it may be true that
human effects are everywhere, it is a matter of degree, and we are
now at a critical juncture in history when we must take great pains
to ensure the survival of those landscapes and species that have not
already been massively manipulated. Open land that has already
been damaged, like these American River canyons, may have to be
restored to membership in the unregulated wild by, for example,
the removal of invasive exotic species and the reintroduction of fire
to the ecosystem.

For me, the bedrock of reality is my affection for wild nature,
and I take exception to the idea that nature is nothing more than a
cultural construction. I do not care if some professor in some rab-
bit warren of a concrete university office building calls my thinking
inexact and sentimental. Sentiment — call it love — for the wild is
ultimately why Will and I became rangers. Sentiment is why any of
us bother to raise children, who sometimes don't appreciate what
we do; why we care tenderly for elderly parents after age has deprived them of the memory of our names. It is why we try to salvage the juvenile delinquent, the alcoholic, the drug addict. Without it we are not human. Perhaps these professors will say that Will
and I lack critical coolness, giving our working lives to protecting

something they say doesn't even exist anymore. In defense, I can only say that to favor a principle — wild, self-willed nature — with the manifest ability to create your species and support you since time immemorial, over a pipe dream of a manufactured and regulated world with no such demonstrated ability, is the most practical thing there is.

11 / EIGHT MILE CURVE

IN THE PREDAWN DARK of a June morning in 1998, a white pickup truck with government license plates pulled onto the dirt shoulder of the Auburn–Foresthill road at a place the locals called Eight Mile Curve, within the boundaries of the land the Bureau of Reclamation had condemned for the long-awaited Auburn Dam. An unmarked government sedan was already parked there when the pickup arrived, and inside it the dark bulk of a person could be made out in the brief flash of headlights of a passing car. Someone got out and unlocked a gate. Both vehicles drove onto a dirt track inside it. The gate was relocked, and the vehicles moved off into the woods.

They had been coming there on and off for months. Sometimes what looked like a civilian vehicle met them there, or a windowless white van with government plates and no markings. But the routine was the same. In a meadow just beyond the locked gate and out of view of the road, the vehicles would stop. Four or five men would get out, greet each other quietly, and begin unloading various gear: white plastic suits and respirators, rolls of red hazard tape like the kind you see at disaster scenes, folding tables, plastic basins marked with warning stickers, bottles of chemicals with hazard markings, and plastic cases like large tackle boxes containing racks of test tubes.

Using waist-high metal stakes, the men would cordon off an area of the clearing with the red disaster tape. Inside it they'd set up the tables, cover them with sheets of disposable plastic, and arrange lab equipment along them. The van would be backed up to the cordoned-off area and its back doors opened to reveal a mobile laboratory inside, equipped with an exhaust hood for handling lethal substances.

Then the men would don their white suits. When they finished they looked like astronauts, their rubber boots carefully sealed with tape to the ankles of their suits, hands sealed in rubber gloves, heads hooded in white, and faces covered by clear masks attached by ribbed hoses to the breathing apparatus on their backs. They could well have been soldiers looking for a missing nuclear warhead or DEA agents about to take down a clandestine drug lab, but they were neither. They were biologists from the state and the county, and for months they had come to this spot in Auburn State Recreation Area to empty traps they'd been setting for rodents.

The routine this particular June morning was a little different. The men set up their equipment but didn't put on their suits right away. As the first hint of pale gray tinged the eastern sky, they walked away from their vehicles into the woods, carrying dark bundles. Ahead of them the Foresthill Divide fell away steeply into the canyon of the North Fork. They set down their bundles and carefully unrolled them. They were lengths of delicate but sturdy netting, made of gossamer threads. The men began fastening them to trees and shrubs. In the pools of light from their flashlights, the woods were unusually green for June. The grass between the oaks was tall, flexible, and dewy. There were wildflowers everywhere, and the cool air was damp and carried the rich scent of April, not the drying-hay odor of the foothills in a typical June.

It had been a rainy winter and the rains had continued halfway into June. In the bottom of the North Fork, below where the men were, the college students I had hired to keep an eye on Lake Clementine were spending their days sitting huddled in their trucks, running the heaters to keep warm. Seasonal workers were required to purchase their own uniforms and fearing they would spend too much of their wages, I had hinted that it was warm in the foothills by Memorial Day, so they might well get through a summer without owning the hundred-dollar jackets. But at 7:30 A.M. when they went to work, the lake was gray and misty, and they froze in their shorts and short-sleeved shirts in the drizzling rain. I encouraged them to take refuge in their pickups and to improvise whatever greenish sweaters or jackets they could until the unseasonable weather ended.

My seasonals were not the only ones who were wet that year. In the early part of 1997 the easterly trade winds that normally push sun-warmed equatorial seawater toward the Asian side of the Pacific, to be replaced along the coasts of North and South America by upwelling of cold nutrient-rich water from the deep ocean, had weakened. Warmer-than-normal water temperatures in the eastern Pacific increased evaporation and cloud formation, driving great wet storms onto the west coasts of the Americas. In Peru torrential rain flooded villages, destroying homes and killing their occupants. The rainwater pooled in low areas, mosquitoes bred in it, and some areas of the country suffered three times the average number of malaria cases. This phenomenon in the Pacific also had far-reaching effects elsewhere in the world. In Kenya and Somalia heavy rains led to outbreaks of waterborne disease, Rift Valley and dengue fevers. But in other areas the 1997–98 Niño had the opposite effect, causing drought, crop failures, and forest fires. My fellow park rangers on Southern California beaches saw unusual numbers of sea lion pups wash up dead in the surf. They looked like rumpled bags of bones. The failure of the nutrient-rich cold upwelling along the coast had led their mothers' prey — squid and small fish — to leave the sea lions' hunting grounds seeking colder water, so the mothers were starving and had little milk for their pups. But on the American River, whitewater rafting outfitters had a banner year, because all the rain and snow in the mountains kept the North Fork at high flows into the beginning of July.

But the men at Eight Mile Curve were not there about El Niño.

When they finished hanging the nets, they extended like invisible fences through the forest, about nine feet high and almost forty long. As the morning light came to the sky between the silhouettes of the oaks, the men quietly retreated into the shadows to wait. Around them the forest awakened in a profusion of birdsong — the plain calls of towhees, the buzzing recitations of Bewick's wrens, the sweet piping of hermit thrushes and Nashville warblers, the raucous squawks of jays. The birds began to flit through the limbs of the trees and along the ground through the underbrush, seeking bugs, grubs, and caterpillars. Some flew into the nets and were entangled. The men emerged from the shadows, gently extricated the frightened birds, and put them in containers, alive and unhurt.

The biologists were employees of the California Department of Health Services and the Placer County Health Department. They had been trapping at Eight Mile Curve since February of the previous year. They had begun by live-trapping dusky-footed woodrats and deer mice. As they would do today with the birds, they had been taking all of the animals to the tables in the cordoned-off work area. When they worked with rodents, no one was allowed into that area without a suit and a respirator, because by now rodents in the Sierra Nevada had occasionally been found infected with bubonic plague and hantavirus, the latter disease previously unknown. These potentially lethal pathogens could be picked up from rodents by humans who handled them.

By this time, creeks in remote wildernesses in the West also contained a human and animal parasite called *Giardia lamblia,* and backpackers now carried high-tech water filtration systems as a matter of course. Such precautions would have seemed ridiculous at the time of my boyhood, when my parents and I traveled widely in the higher elevations of the Sierra, drinking from any creek we pleased. At home my younger brothers and I sometimes found mouse nests in woodpiles and held them wonderingly in our cupped hands, admiring the pink babies with their blind eyes, for none of us had ever heard of hantavirus, or Valley fever, West Nile virus, or Lyme disease.

It was the latter that the biologists at Eight Mile Curve were actually looking for that June day in 1998. Lyme disease hadn't even been identified in the United States until 1975, and no one had heard of Lyme in the Sierra Nevada foothills until the early nineties. Even then it seemed rare and obscure. By that decade, the state Department of Health Services and the Placer County Health Department had began collecting a species of tick that was the local representative of the group that carries the disease to humans and animals, and a fair number of them had tested positive for Lyme. This western black-legged tick, *Ixodes pacificus,* carries the Lyme bacterium — a microscopic corkscrew-shaped thing known as *Borrellia burgdorferi* — in its gut and saliva, and disgorges the bacteria into people and animals when it finishes sucking their blood. Under a powerful microscope the spirochete, as this class of corkscrew-shaped bacteria is known to scientists, looks not unlike that which causes syphilis, and indeed the late-stage neurological effects of the two diseases can be somewhat similar, once the bacteria are fully disseminated in your brain and nervous system. Syphilis, however, is far easier to cure. Lyme is very curable early in the infection but devilishly resistant to antibiotics once it's hidden in the deep reaches of your body.

But the ticks are really just intermediaries, carrying the disease to a new victim after acquiring it from an animal that acts as the bacteria's perennial host. As Lyme research began in California, the perennial host animal — epidemiologists call it the "disease reservoir" — was generally found to be either a dusky-footed woodrat, a kangaroo rat, or one of a couple of species of deer mice. Woodrats are common in the oak forests of the American River. They are far prettier than you might imagine when you hear the word "rat": soft and furry, with delicately colored feet. Their nests, some quite large, are piles of sticks, often stacked against a tree or a rock on sloping ground. These serve generations of rats and their smaller cohabitants, the deer mice.

But at Eight Mile Curve, what the biologists found surprised them. While the ticks they collected on the site tested positive, only one of the many rats and pinyon mice they trapped did. So what was the disease reservoir — the mystery animal that passed the infection to the ticks? At Eight Mile Curve the biologists were seeing a good many songbirds feeding on the ground, and studies elsewhere — Europe, Asia, and the eastern United States — had found Lyme disease in avian hosts. So they turned their attention to birds.

What they then found was published two years later in the *Journal of Medical Entomology:* On the Foresthill Divide, within the lands the Bureau of Reclamation had condemned to build the Auburn Dam, birds were discovered with ticks embedded in them. Many of the ticks carried Lyme disease. Of ninety-two blood samples taken from birds there — towhees, warblers, sparrows, flycatchers, thrushes, finches, jays, and a single hawk — over half tested positive for Lyme; in some species the number was 100 percent. Many of these birds were neotropicals that traveled as widely as flight attendants, appearing in wintering grounds as far away as Central and South America.

Disease on the wing was an idea to get used to in California. Less than three years after the study went to print, the slow wave of West Nile virus, which was by then making its way across the United States at the speed of bird and mosquito flight, reached Southern California. There the virus's arrival was announced, as it had been everywhere else during its cross-country journey, by crows — which, among the avian victims of West Nile, seem to suffer greater mortality, literally falling like sparse black rain from the sky. But even before that, on the spring mornings when those men in moon suits lumbered through the oak forests of the Foresthill Divide carrying containers of little animals, although the air was still full of the familiar, if unseasonable, damp scent of spring, it was a new world.

By 1998 it had been twelve years since I had arrived to work as a ranger on the American River, twenty-three since construction of the Auburn Dam had come to a halt. During an informal meeting with officials from the Bureau of Reclamation in the late nineties, one of our rangers was told, "The Bureau has never deauthorized a dam, and we're not about to start at Auburn." And so our lives in the dam site continued to be a long improvisation. Without a unifying plan for the place — other than its eventual flooding — our effect on it was an aggregate of our individual whims and interests, all expressed in the makeshift way of things that happen under the shadow of a limited budget and a limited future. More often than not the right hand didn't know what the left hand was doing, and although one of my coworkers had made arrangements for the biologists to get into our locked gate at Eight Mile Curve, I never heard about the study until much later. As early as April 1997 the Placer County Health Department was warning citizens in radio ads to be careful about Lyme disease. California State Parks' response to this was not to respond; no cautionary memo was issued to rangers or their seasonal assistants working in the areas where infected ticks had been found. And while biologists who spent only a few hours in our park had taken elaborate precautions against infection, we who were there every day continued to go around in shorts in the summer, our guns no protection against a threat too small to shoot.

Among environmentalists there is a popular fable: If you drop a frog into a shallow dish of boiling water, the frog will fight for its life to jump out; drop the same frog in cool water and then very slowly heat the water to boiling, and the frog will perish with no apparent distress until the last moments, and by then it will be too late. The story refers of course to people's capacity to adjust to increments of strangeness and danger in their environment without taking action to stop it. That may be true of us, but I doubt very much that frogs are that stupid.

I like to mix my own intravenous drugs, making sure that the white powder of the antibiotic is entirely dispersed into the fluid in the clear bag. Once mixed, the plastic bag of fluid is clear and pale yellow, almost exactly the color of urine. I unroll the clear tubing of the infusion set, close the valve, pull the seals on the IV bag and the tubing, and stab the bag with the spike on the end of the tubing. I hang the bag over me from a chrome IV pole on wheels next to the bed. I pinch the clear drip chamber to fill it and then open the valve on the line to purge the air in it. You don't want to get large air bubbles in your brain or lungs. I assemble the other things: syringes, vials of heparin and saline, alcohol swabs, a needle for taking my blood samples. I clean the valve on the end of the tube that goes into my chest and from there to the portal of my heart. I clean it really well, and then I push ten milliliters of normal saline into myself with a syringe. Then I connect the line and open the valve to start the drip. I set the drip at about one per second. I like infusing myself because it reminds me of the satisfactions of competency, of the roaring propwash of a helicopter, of a well-packaged, desperately injured patient on his way into it with me holding the clear bag over him. But now it's more nebulous. I'm not really dying, and I'm not sure day to day if I'm saving myself.

But the feeling in my hands has returned, and I wonderingly run my hands along my nakedness, my chest around the tube that goes right through it into the big veins returning to my heart, my upper arms, the tops of my thigh, which had been numb, too. I imagine the branches of my nervous system, from the trunk of my spine to the tiny rootlets that define the limits of my skin. The sensation of my skin has supplied me with an illusion of a distinct edge, a definable limit between myself and the world. But it's a false autonomy. I know a woman in these mountains who startled and dropped a dish at the exact moment her husband died in a motorcycle crash, miles away. I know a man who carries the canyons of the American River inside himself, in his blood, in his brain.

So many of the things that happened to me as a ranger in the American River canyon I remember well, as my brain gets better. But this one is obscure to me. It was a nothing call that amounted to nothing.

Early June of 1998 was wet. The Douglas firs on the north slopes were festooned with bright beads of water at the time when the crimson Indian pinks were already blooming. Ticks are very sensitive to dryness. Their bodies desiccate easily, and in dry weather they hide deep in mossy crevices on the bark of trees, or in the leaf litter, or in woodrat nests. Damp weather, however, makes them more active, and they come out and stand around on the tips of branches and blades of grass by trails, like hitchhikers at a freeway on-ramp.

June 14, 1998. The dispatch log says that at 1606 hours — six minutes after four — that Sunday afternoon I was on a traffic stop on Lake Clementine Road. What I remember next was hearing a sheriff's deputy radio to his dispatcher that he was pursuing someone on foot down by the river at the Confluence. Someone had done something, but what? Exposed himself to a woman? Threatened someone in a drunken argument? The dispatch log doesn't say, just that it took me seven minutes to get there.

When I arrived at the Confluence, we spent some time chasing this guy through the weeds grown tall in the late rains until we lost him. Then we picked the burrs off our uniforms and left. No one cut any paper. I stayed late at the ranger station catching up on reports and got home around midnight.

The following morning, the first of my days off, I slept late. When I got up I found myself absentmindedly scratching an itch just below my beltline. There was something — a pimple, a little bump. When I finally looked at it, I found a tick — brick red and small, the kind I now recognize as *Ixodes pacificus,* embedded in my skin. I knew enough to pull it out with a pair of tweezers and save it in my refrigerator. I made arrangements to see a doctor, without really thinking anything would happen.

Old Doctor Parsons had an appointment available on Thursday. Like me, he wanted to be a writer, but at the time he was further along in this delusion than I was. His office was a small pale-green building in a stand of ponderosa pine a couple of miles north of Placerville, across the highway from a house with several washing machines and a logging truck in the yard, and about half a mile south of a Scotch and steak roadhouse called the Hanging Tree — one of those places with the chandeliers made of wagon wheels suspended from the smoke-darkened pine ceiling and Freddy Fender and George Jones on the jukebox in the bar. Over the time I knew him, his waiting room became increasingly empty, with a fish tank bubbling in the silence. Eventually he let his secretary go. He was a portly, bald man in his late sixties, and by the time I came in with my tick bite he was spending most of his time hunched over his computer in his back room, cranking out unpublished polemical short stories and libertarian novels in the tradition of Ayn Rand. His writing schedule gave him little time to keep up on developments in medicine.

Lyme disease is rare in California, he said. We think only about one percent of the ticks in the state might carry it.

I handed him the plastic bag with the tick inside.

If we test it and it comes up positive, he said, it'll just make you nervous. Even if the tick's positive, it's unlikely that you'll get the disease. Let's just forget it, shall we?

Okay, I said.

You're likely to get some local irritation and redness, he said. Ticks are dirty animals. Don't worry about it.

And so I didn't. Within a couple of weeks, a round patch of rash circled the bite. Within a month I began to feel very tired. More tired than I had ever been in my life. Then came diarrhea. My hands began to go numb, then my arms, my lips, my tongue, the roof of my mouth. Then my feet. I began to get shooting pains, like hot needles, in my feet.

One day I went to the pistol range to qualify. The police holsters we used had three different safeties to make it hard for someone to take your gun from you in a wrestling match. I couldn't feel any of the releases with my numb fingertips, so I couldn't get my gun out. I stopped going to work. After several months my sick leave ran out.

I was seeing more doctors. A neurologist performed nerve-conductance tests and a spinal tap. Your nerves are damaged and there's protein in your spinal fluid, but I don't know what's wrong, he said. I went home. I fell down a couple of times. My eyes were getting blurry; my ears were ringing. The joints in my fingers and toes were sore and sometimes swollen. I couldn't twist the lid off a jar. I began to go deaf in my left ear. The sound of my young children's laughter cut through me like a knife, rattled my brain. My own speech seemed to reverberate in my face and forehead, each vowel causing excruciating discomfort. My brain felt swollen. I couldn't think. Everything seemed difficult to figure out. I was tired, but I couldn't sleep at night.

I began to lose my memory. I made an appointment with the new chief investigator at headquarters for an interview for a position I had always wanted on the department's investigations team. I never showed up, and it was another week before I knew I hadn't. I missed an appointment with my dentist, and then another. I rescheduled and missed it again. The dentist's secretary told me it might be time to find another dentist. I called up the dentist, a friend, and found myself weeping on the phone. I didn't know myself anymore.

I was tested again. I saw a fifth doctor, a sixth, and then a seventh. I was told I had Lyme disease. I tested positive. One day two years after the bite I drove down to the ranger station, a route I had taken for thirteen years. I was now feeling drugged or drunk most of the time; my brain was full of spirochetes, a doctor told me later, and the inflammatory nature of the body's reaction to them causes swelling in the walls of small arteries, resulting in decreased blood flow to the brain. When I finished my errand at the ranger station, I started home again. But I wasn't sure anymore exactly how to get there. I pulled over to the side of the road and called my wife on my cell phone, then sat there waiting for her to come and get me. I waited for months for things to get better. Eventually I was forced to retire from State Parks, and for two years I wasn't much good for anything. The trail back from that bad place was so long and circuitous that, like many Lyme patients, I cannot say exactly when it began to look like I was going to get better. As I write this, it isn't over yet.

However, as I healed, I made two very pleasant discoveries. One was that my memories of what had happened during all those years in the American River were intact. It was as if they and so many other things I was missing — the names of friends, phone numbers, parts of my vocabulary, and the contents of books I had read — had been locked in a cabinet during my illness, and now I had the key again. The other surprise was that as I was healing, the American River's situation had gotten better, too.

In the final analysis, neither of my present doctors can promise me that the Lyme spirochetes or their cyst form — like a seed — will ever be completely absent from my body. But then, when you think about it, who ever thought that a ranger could spend fourteen years on a piece of land and the two would remain entirely separate? Environmentalists have been saying for years that as the land goes, so will we go. It should be no surprise to learn that rangers may be the first to know how true that really is.

EPILOGUE

IF FOR DECADES we park rangers risked our skins to make the American River canyons safe for river lovers to visit, it was the river lovers who kept them from going underwater, and who finally brought a symbolic end to the Auburn Dam with the same earth-moving machinery and concrete that had been used to partially build it. Here is how that happened.

In the late 1960s, when the Bureau had begun work on the dam and Placer County's Middle Fork dams had just been completed upstream, a tunnel nearly three miles long was blasted from the dam site on the American River to the next drainage north, Auburn Ravine. Had the Auburn Dam been finished, the Placer County Water Agency could have released stored snowmelt from their Middle Fork dams into the Auburn Reservoir and withdrawn it again through this tunnel, by gravity, to supply the western edge of the county with its water.

Three decades later, the mouth of the Auburn Ravine Tunnel — which by then should have been four hundred feet beneath the Auburn Reservoir — was still high and dry on the canyon wall in the dam site, two hundred feet above the river. Meanwhile Placer County's population had tripled, much of that along the county's western edge. To fulfill its obligations under a contract made with the county's water agency back when Auburn Dam was a certainty, the Bureau had been installing a temporary pump station and pipeline from the American River up the canyon wall to the mouth of the Auburn Ravine Tunnel every summer. Every fall the agency would disassemble the whole affair and move it to high ground, because the pumps had to be located where the river entered the diversion tunnel around the dam's foundations, and the shape of the canyon there made the pumps and pipeline vulnerable to being swept away by high water in the winter. This was costing the federal government between a quarter million and, on at least one occasion, one million dollars a year.

Over time this situation made strange bedfellows. Even with the temporary pumps, Placer County was getting less than a third of its annual rights to American River water at the dam site. For its part, the Bureau wanted an end to the Sisyphean task of constructing and disassembling the pump station and pipeline. Until such a time as the Auburn Dam's political fortunes got better, the way out for the Bureau was to close the diversion tunnel, restore the river to its course through the dam site, and install a permanent pump station in what was now the dry part of the riverbed, where the canyon's shape would put the pumps out of reach of high water. And environmental groups had long seen restoration of the river through the dam site as the victory they wanted over the Bureau and its dam.

In 1995 the Bureau quietly began working on plans to do what everyone wanted. But by 1998 the project ran afoul of Congressman John Doolittle, now chair of the Subcommittee on Water and Power Resources and a member of the powerful House Appropriations Committee. Doolittle saw the restoration of the river through the dam site the same way the conservationists did — as a symbolic end to the Auburn Dam — and he didn't want any part of it. In his position of power over the Bureau's budget he demanded that the Bureau redesign the project to put the pumps somewhere upstream, leaving the diversion tunnel and other completed work on the Auburn Dam intact. However, if the Bureau did this, the project would lose the support of environmentalists.

A book such as this can contain only the highlights of such a long and convoluted story as the Auburn Dam's, in which two generations of conservationists fought two generations of developers, politicians, and dam engineers, and enough studies and reports were published to fill a medium-sized library. But to sketch that story's resolution, I must mention just a few of the environmentalists who had a part in saving the American River, and to do so should not be seen as diminishing the contributions of others.

Gary Estes is a slight, quiet, conservative-looking man who favors slacks and button-down sport shirts in muted designs. He has a degree in political science and the remnants of a soft Virginia accent. His professorial appearance and speech belie his blue-collar origins. His father was a ship fitter at the Norfolk Island Naval Shipyard.

Estes's résumé is a strange one. He is probably one of those people with a very high IQ who can easily become bored with the mundane quality of everyday work in most fields. By the time his restless intelligence found the Auburn Dam, he had already been a schoolteacher, a self-taught engineer specializing in heating and cooling of office buildings, the administrator of an energy users' group, an activist in energy issues, and a paralegal. In 1989 Estes and his wife moved to Auburn and Estes became his own general contractor, constructing a house for himself and his wife on the rim of the North Fork canyon near the dam site. By the time he finished the house he'd become interested in the dam. His wife had become used to being the breadwinner while Estes was building, and she now offered to support them while Estes devoted himself to his wide-ranging avocations. Estes is the first to admit that makes him a lucky guy.

In his new time off, Estes decided to teach himself geology and seismology in order to review a mountain of technical literature on earthquake hazards at the Auburn Dam site. He then prepared a treatise on the subject to accompany a presentation to the Auburn City Council. In his presentation, Estes pointed out that constructing the Auburn Dam to prevent Sacramento floods merely shifted disaster risk from the people of Sacramento to the people of Auburn, who lived next to the faults along which the Auburn Dam's filling might trigger a devastating earthquake. However speculative, it was an interesting point.

During the same period Estes was reading advanced meteorology, climatology, and river hydrology in order to study the kind of storms that caused floods on the American River. Estes thought it might be possible to recognize these storms as they formed over the Pacific (as Bill Mork and his colleagues had in 1986) and then convince the Bureau and the Army Corps of Engineers to act on those predictions (which did not occur in 1986) by dumping water from Folsom Dam to accommodate the predicted inflow. This, he reasoned, might eliminate the need for an Auburn Dam — at least where the dam's flood control aspects were concerned. In 1994 Estes coauthored a scholarly paper on the subject with a local college professor and then founded an annual conference at which to present it. A decade later the California Extreme Precipitation Symposium continues to attract top experts in weather, climatology, and flood control from all over the country, and Estes is still its principal organizer and moderator. The symposium has been credited with fostering an exchange of ideas on ways of using weather forecasting and storm modeling to better operate dozens of flood control reservoirs throughout California.

By this time Estes had joined a tiny and virtually penniless Auburn anti-dam group calling itself Protect American River Canyons (PARC). He soon became PARC's resident technical wonk. His trespasser's intellect gave him an ability to digest voluminous reports in other people's disciplines, and his lack of steady employment made him reliable about showing up at public meetings. In May 1997 Estes and fellow PARC activists attended a meeting at which the Bureau made a presentation on its plans to restore the river at the Auburn Dam site. Estes left feeling excited. But two years later, with Doolittle's redesign under way and the tripartite support for the project in danger of losing one of its legs, the project had slowed to a crawl.

Estes the former paralegal now took up lawyering. Poking around in law libraries for something with which to dislodge Doolittle, Estes discovered the public trust doctrine. Simply stated, the doctrine dictated that no one had a right to keep the public from using a navigable waterway without demonstrating just cause or exigency. Estes reasoned that the American River was navigable by kayak and raft, but by virtue of the danger that whitewater boaters could be sucked into the diversion tunnel and drowned, the portion of the river through the dam site had been closed to them for thirty years. Yet no dam had been built to justify that exclusion. Bingo, thought Estes.

Protect American River Canyons' larger ally in the fight against Auburn Dam was the Sacramento environmental group Friends of the River. FOR employed a man by the name of Ron Stork. With his short, curly gray hair, wire-rimmed glasses, and button-down oxford cloth shirts, Stork looked less like an environmentalist than an aide to Doolittle or some other Republican legislator. By the early nineties he was well known as one of the deadliest weapons in the environmentalists' arsenal against dams. Stork shared Estes's facility with facts, figures, and technical language; he was a good public speaker and strategic political thinker; he was single and without children; and he was possessed of a mighty work ethic that often kept him late at the office. In 1992 Stork spent four months camped out in a Washington, D.C., home converted, with seven cots and a couple of couches, into a dormitory for environmental lobbyists. From there he and his fellow campers walked back and forth on Constitution Avenue to petition Congress successfully to defeat that year's Auburn Dam bill. He did the same thing again in 1996, and this time, for a few weeks, Estes joined him. The two men became friends. Always humble, Estes thought of Stork as the *über* Estes.

At the center of PARC were two tall, lanky men: Eric Peach and Frank Olrich. Peach had an upholstery shop in Auburn, and he was so tall he had to build a platform on which to put the chairs he reupholstered, so he wouldn't have to bend over. Olrich, an education consultant, had been a high school basketball star in Auburn. By 1998 Estes was discussing public trust doctrine with Olrich and Stork. The three had quietly begun consulting experts preparatory to suing the Bureau to force the latter to restore the river. Their first consultation was with a Sacramento lawyer who was an expert in the public trust doctrine. Encouraged by this meeting, Olrich placed a secret call to a Bureau official in Colorado for advice. The Bureau man's hobby happened to be kayaking, and the two men had become friends when they met while running a river in the Rockies.

"Off the record," the Bureau official told Olrich during this call, "you should sue us for access to the dam site under the public trust doctrine. You'll probably win."

At this point PARC and FOR suffered setbacks. First they couldn't come up with money to mount a major environmental lawsuit. Then, in the winter of 1999, Olrich died while cross-country skiing with his wife. Making their way through deep snow in the mountains east of Auburn, they attempted to cross the trans-mountain railroad tracks by climbing through the deep trough left by the trains' snowplows. Olrich was unable to climb out in time to avoid being struck by a freight train. His wife survived.

The funeral was held on a cold, windy February day at the old cemetery above the railroad tracks in Auburn. A Native American shaman chanted over an open mahogany coffin in which Olrich's mangled remains had been covered with a mixture of cedar shavings and dried flowers. Dark-eyed juncos and white-crowned sparrows flocked through the graveyard around the mourners huddled in their overcoats and mountain parkas, as if coming to pay their last respects. The ceremony stopped briefly when a Union Pacific freight like the one that had killed Olrich thundered up the valley, drowning out the eulogy.

Meanwhile a new Democratic administration had taken over the California governorship. Searching for a deeper pocket to back their effort, Stork and Estes approached the state attorney general's office to see if it might want to be a party to a lawsuit; after all, the dam site was a state recreation area. "Yes," said their contact at the AG's office. "We're interested. Let's have a meeting."

When Stork and Estes walked into that meeting, they were astounded to have the very lawyer they'd been consulting on public trust doctrine introduced to them as the new acting chief of the attorney general's environmental law branch. Within weeks the Bureau of Reclamation received a letter from the attorney general saying the state had decided to sue if the Bureau didn't move forward with restoring the river through the dam site and allowing the public to use it. The Bureau's lawyers evidently thought the state could win, so the process was set in motion to put the river back in its course.

For a while John Doolittle continued to blockade any federal legislation to improve flood control for Sacramento that wasn't an Auburn dam. But eventually, with no dam in sight, he came to an agreement with Sacramento congressman Robert Matsui to back improvements to levees in Sacramento and modifications to Folsom Dam, which would give the city protection from any flood large enough to have less than a one in two hundred chance of occurring in any given year.

By 2003 the Auburn Dam site resounded with the roar of earthmoving equipment as contractors took down remaining portions of the cofferdam, prepared to close the diversion tunnel, and began installing a manmade rapid with cemented-in boulders and a complex system of manifolds to pull Placer County's water out of the river without sucking up kayakers, protected fish, or the huge cargo of rock and mud the river carries at high water — a feat of engineering that, as far as anyone knew, had never been done anywhere in the world. Stork and Estes were as ecstatic as such quiet, studious men ever get. Some members of PARC had been suggesting that State Parks name the restored stretch of river the Frank Olrich Confluence Parkway. So far State Parks has not said it will do it. But it is safe to say that in the short run the future looks bright for the American River.

The river's long-range future is less certain. By 2000 an eminent panel of experts from government and top universities reported that the complex human-caused climatologic changes for which the term "global warming" had for some time been an inadequate handle was well under way. Most other studies concurred.

Among the effects forecast for California by the 2000 study was a generalized increase in total precipitation; others disagreed, saying California would become dryer. But there is general consensus that warming temperatures will raise the altitude at which rain becomes snow on the mountains, and as a result more precipitation will fall as rain and run off immediately, instead of sitting in the snow pack and percolating off the mountains during the spring months. Further, studies agree that winter rains will arrive less reliably, and that as the weather's heat engine revs up, there will be an increase in warm, high-energy events like the Pineapple Express storm of 1986. If the second half of the twentieth century is any indicator this may be true; these storms seemed to grow larger between 1955 and 1997. Such storms would raise rivers to flood stage more often, and the floodwaters would have to make their way to the sea through a Central Valley with ever greater population and property values. Turning from the question of floods to that of water supply, a diminished snow pack melting earlier in the year and more precipitation running off as rain would give agriculture, cities, and industry a smaller annual period during which to gather their water, so they may demand more space in which to store it. Environmentalists hope that groundwater storage may to some extent offset the need for more dams.

Most studies agree that under current climate change scenarios, droughts and heat waves will lead to greater demand for hydroelectric and other power to run air conditioning and higher water use for landscaping and crops. At this writing California has not had a serious drought for a decade and the two worst dry spells in the twentieth century were only six years long. Yet studies of ancient tree stumps found on lake bottoms show that California experienced a two-century drought at the end of the first millennium and another one nearly as long around A.D. 1200. And for the present and foreseeable future — unless something changes — the forty-two-thousand-acre dam site called Auburn State Recreation Area will continue to be owned by a government agency whose purpose has always been to develop water storage. The writer John McPhee visited the Auburn Dam site in the course of his research for his book about geology, *Assembling California*. While there, he learned about the timber dam that had existed on the site as early as 1854. McPhee would remark dryly, "[At Auburn,] environmentalists have discovered to their eternal chagrin, a dam site is a dam site forever, no matter what the state or nation may decide to do about it in any given era."

Nevertheless, the American River canyons are resurgent and bursting with life. In 2001 the city of Auburn — a freeway burg just over half an hour by car from the capital city of the most populous state in the union — had a new problem. Bears coming out of the American River canyons at night had been rampaging through the streets, knocking over garbage cans and startling residents awake. By 2003 Auburn was being fortified with a firebreak along the canyon rim to defend the town from brushfires that periodically sweep the canyon's fire-adapted chaparral. Deer, foxes, and bobcats had been leaving footprints in the dust of the firebreak as they joined the bears in the streets of town. Biologists conducting a reconnaissance of the North Fork had found legions of foothill yellow-legged frogs, *Rana boylii*, whose decline elsewhere in the Sierra had been worrying herpetologists for some time. And Lake Clementine Road, which runs from just above the Auburn Dam's waterline to four hundred feet beneath it, appeared in a guidebook of the best spots to see wildflowers in the Sierra Nevada.

A couple of months before his death, Frank Olrich had come to the ranger station with an idea about forming a group of volunteer docents he called the Canyon Keepers, who would offer the sort of nature walks that the beleaguered rangers didn't have time to do. Today, Canyon Keepers is an official state park volunteer association, and as many as sixty people have been showing up for some of its guided hikes. Meanwhile, up Foresthill Road, a federal grant has brought dignified paved parking lots to the formerly dusty trail-heads. Local pride in the American River is growing. The *Auburn Journal* now runs scenic photographs of the river on its front page as often as two or three times a week. Some of the park's signs don't even have bullet holes in them anymore. And the crusty all-male ranger staff has been leavened by the arrival of female rangers and a woman superintendent.

In 2000, Bruce Kranz, State Parks' superintendent for the American River, ran for a seat on the Placer County Board of Supervisors against a moderate, controlled-growth, environmentalist incumbent. Kranz's campaign literature claimed that as an employee of State Parks, he'd devoted himself since the age of seventeen to the stewardship of California's environment; however, builders and developers were prominent among his campaign contributors. On Election Day, when it was too late for the press to report it to the voters, his campaign received $50,000 from a political action committee tied to Congressman Doolittle and development interests. Apparently the voters didn't entirely buy Kranz's credentials as a green; he outspent his opponent by a wide margin and was defeated by a much narrower one.

Increasingly isolated among the preservationists at State Parks, Kranz retired from Parks to devote himself to politics, taking a part-time job as an aide to conservative state senator Tim Leslie. In this capacity, in 2002, Kranz appeared at an outdoor press conference overlooking the foundations of the Auburn Dam to deliver a speech in support of a new effort by Leslie, Doolittle, and fellow conservative legislators to finish the dam and flood the landscape he had stewarded. The following year he ran again for county supervisor. This time he successfully ousted the incumbent with a generous advertising budget. Kranz's campaign raised $241,000 from a list of contributors that reads like a directory of the building and land development industries in the Sierra and Sacramento regions. His controlled-growth opponent raised less than half that amount.

By this time Steve MacGaff had retired. O'Leary retired, went fishing in British Columbia, and settled close to the coast. Doug Bell retired, went hunting and fishing, and now designs hiking trails. His daughter is studying ballet at a good university. Sherm Jeffries is a superintendent for State Parks on the east side of the Sierra.

Only Dave Finch and Will Reich still don their uniforms each morning in the old firefighters' mess in the North Fork. Finch retired once but couldn't stay away. Reich still patrols the North and Middle Forks in an oar boat. There is still no word of the miner who was run out of his claim at Otter Creek by a bear.

After the beating of Ricky Marks by Mary Murphy's boyfriend, Marks and his partner, Jerry Prentice, abandoned their cabin on the North Fork. The rangers burned it down in the winter of 1987. Marks and Prentice continued mining for a couple of years but never struck it rich. Eventually both men left the Sierra for cities on the coast, where Marks was later arrested for drunk driving. Out on bail, he failed to appear in court. A warrant has been issued for his arrest. Mary Murphy left her abusive boyfriend and moved to New Mexico, where she completed a twelve-step program and now lives happily and peacefully without alcohol or alcoholics.

New road maps published by the California Automobile Association no longer show a Y-shaped lake in the canyons of the American River.

Each year on July 11, Early Ditsavong's mother and father return to the place where the North Fork runs into Lake Clementine with offerings of incense, flowers, ceremonial food, and a candle to light Early's way through the spirit world. In 2002 the California National Guard bestowed its highest posthumous honor, the Medal of Valor, on Private First Class Early Ditsavong for the bravery he displayed in the rescue that cost him his life. He had been scheduled to go to army basic training the week he drowned. I never again saw anything like that lady beetle migration I encountered the day I took Early's parents to the North Fork. Nor has there been another earthquake of the size of 1975's on the Foothill Fault Zone, although we are undoubtedly due for one. There has not been another mountain lion attack in Auburn State Recreation Area in the decade since the death of Barbara Schoener. And Karen Dellasandro's body still has not been found.

In April of 1990, John Carta made another parachute jump off the Foresthill Bridge — this one without a motorcycle. When he landed, he hid his parachute in a camouflaged bag under the span, intending to retrieve it after the rangers and sheriff's deputies went home. It was discovered by the rangers and seized as evidence. That summer, using another rig in a jump from a building, he broke his back. Still wearing plaster casts, that September he accepted an invitation for an airplane ride over Clear Lake, across the Central Valley from Auburn. The inexperienced pilot put the two-engine aircraft through a series of showy, low-altitude maneuvers, then lost control and crashed into the lake. Carta, the pilot, and six others aboard were killed instantly.

In September of 2003 festivities were held to commemorate the thirtieth anniversary of the Foresthill Bridge's dedication; Placer County officials were still debating what to do about suicides there. I understand that recently a marriage was conducted at the bridge, after which the bride and groom parachuted into the North Fork canyon.

By the time this book is in print, it will have been four decades since the legislation authorizing the Auburn Dam was passed, and the forty-eight miles of canyons the dam would have inundated are still with us. Given half a chance, these canyons will continue to bloom and recover from the insults of the Gold Rush, yet they will henceforth lack the feeling of permanence they must have had for the first people who knew them. What has happened to them has rendered these canyons mortal in our eyes, and like the rest of wild nature, they will now continue to exist only at our sufferance. It is my hope that there will be rangers watching over them for a long time to come, or at least until armed men and women are no longer necessary to protect such places.

ACKNOWLEDGMENTS

I OWE SO MUCH TO MY EDITOR, Deanne Urmy, who helped me to shape this strange little book from the very beginning. The steadfast support of my agent, Sandra Dijkstra, and her staff too have been a source of great comfort. Manuscript editor Beth Burleigh Fuller brought new energy and a fresh point of view to our work, late in the editing process.

Gary Snyder nudged me toward telling this story when, at a party one night in the early 1990s, he asked me how my work as a ranger was going. I told him that working under the waterline of the Auburn Dam bothered me terribly. "That seems like a good subject for an essay," replied Snyder, and in my suggestible state as an aspiring writer in the presence of a master, I set out to write one. Casey Walker edited that first essay for the *Wild Duck Review,* and Emerson Blake edited the next one for *Orion.* I am thankful for their insightful suggestions. My special thanks to Nina Leopold Bradley, daughter of Aldo Leopold, who wrote me when the first essay came out in print, encouraging me to continue.

My thanks to the geologist Richard Hilton and the botanist Joe Medieros for their friendship, encouragement, and assistance, and for the love they've instilled in the students of Sierra College for the mountains for which their school is named. California's state climatologist Bill Mork and hydrologist Maury Roos of the Department of Water Resources gave me invaluable help in their fields. Naturalist and writer David Lukas showed me new wonders on our hikes together in the American River canyons, which I thought I knew well before. So many others enriched various fields of knowledge I needed to understand what I was writing about better: Michael Barbour on Mediterranean vegetation; Jeff Mount on river geomorphology; Mike Lynch on the history of the American River; Lucia Hui, Stan Wright, and Mark Miller on ticks and Lyme disease; Kevin Hansen on mountain lions; and Mike Catino and Tom Aiken on the history of the Auburn Dam. Sharon DiLorenzo helped me find another key witness to that history. Jill Dampier and Nick Willick assisted me in my search for old records. Mark Hewitt, Robin Heid, and Matt Davies taught me about the esoteric sport of parachuting from bridges. I benefited greatly from an unpublished multivolume compendium of the history of the American River by John Plimpton. The work of Norris Hundley, Donald Kelley, Robert Pisani, and Gray Brechin also proved indispensable.

For their encouragement, mentoring, generosity, and friendship, I will always be grateful to Scott Russell Sanders, Wendell Berry, John Hart, Terry Tempest Williams, Barry Lopez, Robert Michael Pyle, Rebecca Solnit, George Sessions, Patianne Rogers, Alison Deming, Louis and Brett Jones, Oakley and Barbara Hall, and Michael Carlisle. I am also deeply thankful to Marion and Olivia Gilliam, Aina Barton, Laurie Lane-Zucker, Jennifer Sahn, and the staff and board of trustees of the Orion Society for their early and longtime support of my work.

To my wife, Susan, and my children, James and Emma, I extend my deep gratitude for their love and wholehearted support.

My father, John Fisher-Smith, learned that my mother was pregnant with me on a backpacking trip in the Grand Canyon. Later he and my mother led my brothers and me into the wilderness, patiently fanning sparks of love for the mountains and canyons into flame. That fire has never gone out, and for that I am deeply thankful. Stephen Studebaker, by turns a schoolteacher in the Navajo Nation, a railroad conductor, and a seasonal park ranger, inspired me to take up rangering. During a climbing trip together in the High Sierra thirty years ago, he wore a park service uniform and got paid, and I didn't, and didn't. To my youthful eyes it looked like a good racket. It turned out to be more like work than I had imagined. Mike Whitfield and John Kraushaar taught me much about my job in the early years. Inspector Brian Dressler and Sensei Rod Sanford of Zen Bai Butoko-kai and the Pacific Institute of Defensive Tactics taught me much of what I know about being a law enforcement officer. To Mike Van Hook, my apologies and my thanks for putting up with me.

Perhaps most of all, my thanks to every ranger I ever worked with, the many more I didn't, and those who follow me. You continue to stand in defense of the sweetest and most hopeful places I know: the world's national, provincial, and state parks and wildernesses.